Against Technoableism

NORTON
SHORTS

Against Technoableism

Rethinking Who Needs Improvement

ASHLEY SHEW

W. W. NORTON & COMPANY
Celebrating a Century of Independent Publishing

For information about permission to reproduce selections from this book, write to Permissions, W. W. Norton & Company, Inc., 500 Fifth Avenue, New York, NY 10110

For information about special discounts for bulk purchases, please contact W. W. Norton Special Sales at specialsales@wwnorton.com or 800-233-4830

Manufacturing by Lakeside Book Company
Book design by Ellen Cipriano
Production manager: Delaney Adams

ISBN: 978-1-324-03666-1

W. W. Norton & Company, Inc.
500 Fifth Avenue, New York, N.Y. 10110
www.wwnorton.com

W. W. Norton & Company Ltd.
15 Carlisle Street, London W1D 3BS

1 2 3 4 5 6 7 8 9 0

To disabled everything, and everyone,
especially my local crip community in the
Disability Alliance and Caucus and the
New River Valley Disability Resource Center

CONTENTS

Against
Technoableism

CHAPTER 1

Disabled Everything

▪　　▪　　▪　　▪　　▪　　▪　　▪

L ET'S START THE WAY YOU expect a good disabled memoir or
crip story to start: either with my horrible birth or my grave-brave
injury. It could be my monstrous birth—a real term in medical history,
by the way—where I was born and had to fight against great odds even
as an infant, and my story could end in triumph where I . . . continue to
live and am writing a book about it. Or I could start with a freak accident that helped me steel myself against this cruel world, or even better,
some gnarly war story: my heroism and sacrifice, the pledge I made to
my brothers (siblings) in arms, living so their deaths will not have been
in vain, paying the toll of freedom with my body. Then I could show the
ableds how to *really* live (and laugh, and love).

Ooooh, let's really lean into these narratives, mix up the stories
for some real pizzazz: I'm born disabled, no one thought I would live,
but then I become of military age, and I use a machete to fight for
your freedom! So young, so heroic, so very disabled. You could meme
images of me with the words "What's your excuse?" superimposed
over my twisted body swinging that machete as I sit atop DARPA-

developed disability technology, a super-duper supersoldier with an inspiring message of vanquishing one's enemies and one's disabilities with the power of technology. Or maybe I could just die young so *you* can tell my story. I won't sully my own good name by disappointing you with my normalcy. You could talk about how you learned so much from me or say how I'm no longer in my earthly prison (by which you mean my disabled body and mind). You can imagine me "whole" again (as if being disabled broke me, even though that's how I was born or how I live my life). My parents, if I have them, can pen the story of how they learned to love me despite my many challenges; they will, of course, have to share their story of grief about my existence. Is this the disabled drama you want?

This kind of story might be what you expect, or even desire, but it's not what you're going to get. Disabled people are often asked to curate their stories to frontload their dramatic origins: I don't begrudge those who start this way. But I like the poetry of The Cyborg Jillian Weise, especially the one where Cy (our disabled cyborg laureate) ventriloquizes a nondisabled audience member in "Nondisabled Demands." The nondisableds have dragged our dear cyborg up to the stage, and instead of asking about poetry or any area of expertise, they demand Cy's disability origin story.

> *You can't expect people to read you*
> *if you don't come out and say it.*

Before they will listen to the rest, they need the story, need to find out how inspired they should be. I see this same thing happen over and over: disabled experts and professionals are continually getting asked inappropriate personal questions when they are there to offer their professional expertise. They are not treated like other authors/scientists/ social workers/teachers. Sometimes their expertise is rooted in their personal history (whose isn't?), and a personal narrative is fine when they offer it, but sometimes they are treated as a sideshow—a curiosity added to an event but taken less seriously than other experts. I see this play out most often for my autistic colleagues who have *professional* expertise in psychology, medicine, social science, human development, and education. They are asked to be part of events about autism or neurodiversity, but they aren't treated the same as other participants, and they are very rarely paid the same (if they are paid at all).

While I hate that disabled people are forced to tell their stories, I do love stories. In fact, parts of this book will involve some of my own tales. I just don't think most disability *origin* stories are very interesting. Mine is certainly not. Some *are*, don't get me wrong. There's a guy in one of my amputee groups who had his leg ripped off in a grain auger accident, and he did not pass out while he waited for emergency rescue. It's always a shocker for newbies in my amputee groups who have come with stories less gnarly. I love when he shares his story with obvious relish: he's got it well rehearsed for other amputees.

This kind of sharing among those in the community is different, though: the lessons are different. It's usually not about being on the other side of something we've overcome. We survived or were simply born that way, some of us also have PTSD, but we also have normal lives and do average things. We ask each other about how kids adjusted, how family members reacted. There's commiseration about dating and jobs. The improvements we see are incremental adjustments, sometimes nonlinear, with more surgeries needed, other conditions to manage, insurance woes, equipment issues. We compare prosthetic feet and offer ideas about physical therapy. We don't all agree about how to approach situations, nor about what feet work best, nor about dating for that matter. Things aren't tidy or managed. I like when it's just us, and I like when the stories aren't crafted for nondisabled expectations, like so many disability stories are.

THIS IS A BOOK about the stories that disabled people tell that nondisabled people usually aren't interested in. I am a disabled person who uses various technologies. But my thoughts on technology aren't the whole story. I don't use all the technologies out there aimed at my disabilities (it's hard enough to get some that I do use), and I certainly don't claim to represent all disability communities—or to be a representation of any sort, because we are all individuals. Instead, I intend to take seriously the experiences of disabled people in a world that often considers technology the "solution" to the "problem" of disability. The scare quotes indicate real fear here—and I hope this book will help you understand why.

When people posit disability as a *problem*, they look for solutions. Disabled people can and do have problems, which sometimes include pain and dysfunction. However, many of our problems are social, structural, and practical problems that stem from the idea that disabled people are fundamentally flawed, unworthy of inclusion, broken, or inadequate. This is ableist thinking. We have to resist these kinds of assumptions, which produce simplistic disability stories that collapse disabled people into stereotypes.

When I was in grad school, before I was disabled, I made great friends with my fellow student Jill. I didn't think of myself as disabled, and it would be another ten years before I would be diagnosed as having Crohn's disease, but I was already having intermittent issues with inflammation and flares of symptoms that didn't really make sense together yet. She was a paraplegic wheelchair user. She passed away over a decade ago now, but back then, we hung out with a larger group from our cohort, and Jill and I got along especially well. We had classes together, occasionally shopped together, and she'd invite us all to her apartment for group hangs. She was originally from New Jersey but had moved to New Mexico, where the climate was better for her. She loved New Mexico. She was just away to earn a master's degree: two years in our program in Virginia, then she planned to return home to New Mexico to teach. She was full of jokes, and it was easy to be with her.

On our campus, disability accessibility was . . . shitty, to say the least. And she was the first physically disabled person whose experience I could really see clearly every day, since we were in so many activities together. I remember the long routes we had to take to avoid stairs. I was lucky to be in a ground-floor apartment, so she could occasionally visit where I lived, though the bathroom door was too narrow, so visits weren't terribly long. Her own home wasn't ideal either. She had very little choice over what apartment she could move into—so many fewer options than I had. And the apartment she ended up in wasn't even truly accessible. It had a bathroom with an appropriate sink and toilet, but the shower wasn't great, and the kitchen was just the same standard kitchen that was in all the apartments in the complex, so she had to be creative. The apartment complex itself was kind of a dump—filled with undergraduate students and close to campus. The parking lot was always overfilled on football game days. On one of those days, we were going to hang out, and then she was going to go over to the library for an event. I drove over, found a parking spot way out in some grass, and trekked over to her apartment.

Jill was on the phone when I got there. Someone had parked next to her giant white van in the yellow-slashed area that is the open zone next to her disability parking spot. She could neither get to the side of her van to get in nor open the door to allow the van's wheelchair lift to fold out. The van was set up for her, with no seat on the driver's side since she'd be in her wheelchair, so I couldn't simply back it out for her to get in. (It was set up for someone like her, but not her, actually. She considered herself very lucky to get a secondhand wheelchair-accessible van like this; it had been set up and driven by a different person with a similar disability. A new van would have been way out of a reasonable range of cost, and the secondhand van was still pricey.) She had already called the apartment complex once that day. Although the complex had signs all around the lot saying that it would tow away non-resident cars (which this one was!), it was game day, so they didn't want to call a tow company. They said they couldn't do anything and told her to call the police. She called the non-emergency police line, and they said it was the domain of her apartment complex. And it was game day after all, so the police were busy too. She was shuttled back and forth like a tennis ball. No one was going to do anything.

I was outraged. Why should my friend be kept home all day because someone decided to park where they shouldn't? What if there had been an emergency of some sort for which she'd needed her van? Didn't they know that disabled people had places to go?

Jill wasn't pleased either, of course, but this wasn't her first time hitting walls with access. I was naïve to be so mad. She encouraged me to let the air out of the car's tires. I regret that I was too much of a wuss at age twenty-three to actually do it. We already knew the police weren't going to come about the car, so why didn't I? *Often, past me disappoints my currently disabled self.*

I think of all the infrastructure that failed here—that was really built to fail, because it was never built to consider disabled people. It did the bare minimum: the physical complex (and administrators, and the police) met a few legally required standards, but they did not actually enforce the civil-rights law that is the Americans with Disabilities Act (ADA). Few complexes met even the lowest bar, so her housing options were limited—she was glad to even get that dumpy, inconvenient apartment. The systems in place supposedly to give her equal access—including her leasing office and the police—were indifferent to actually ensuring that she had it. Jill missed the event she said she'd help with at our library. Turned out she was accustomed to these system failures. At least that day wasn't a failure with life-or-death stakes, like the ones that would delay her access to gynecological care and ultimately lead to her death less than two years later.

One of my disability twitter follows, @MsSineNomine (Gabrielle Peters), used this phrase a few years ago: "Make it accessible or burn it down." It's since been printed on shirts,[1] and I painted the words onto a canvas and hung it in my office . . . which is on the first floor of a three-story building without any elevator, away from all my colleagues, where I was moved when I became disabled. Where every doorknob is non-compliant with the ADA. I'm not plotting arson. But like Jill, I'm forced to think about inaccessibility a lot. Sometimes the stakes are our lives—our literal survival, meaning that we could die due to inaccessibility and shoddy emergency planning. Sometimes the stakes are our lives—our social, work, family, and love lives, meaning that we can be prevented from physically inhabiting places, going and doing things, having relationships and friends, doing meaningful work in the world. You know, having a life.

I CAME UP WITH the term "technoableism" to describe a pattern disabled people see over and over—and a pattern observed by many others too.[2] Technoableism is a particular type of ableism, one that is highly visible in media and entertainment and omnipresent in the ways most people casually talk about technologies aimed at disability. Technologies for disability can never just be "tools that are useful sometimes," in the phrasing of Jen Lee Reeves. Technoableism is a belief in the power of technology that considers the *elimination of disability* a good thing, something we should strive for. It's a classic form of ableism— bias against disabled people, bias in favor of nondisabled ways of life.[3] Technoableism is the use of technologies to reassert those biases, often under the guise of empowerment.

Disabled people are often on the front lines of tech users; we are considered the first testing ground for new tech like exoskeletons, Closed Captioning, automatic door openers, text-to-voice and voice-to-text software, and more—tech that is then picked up by all. Yet we are usually left out of the conversations around tech for disability. There are ongoing calls for universal design, but this design employs and imagines our bodies (often as the basis for professional certification) while rarely (if ever) simply asking us what we need. "Engineering for good" labs design prosthetic hands without ever talking to hand amputees. The public is always ready to ask us why we don't have whatever hot device has recently been hyped in a feel-good news story. While our lives are deeply entangled with technologies of all kinds (not just fancy ones: left-handed scissors or walkers or hearing aids are all disability technologies), disabled people are almost never included in discussions about what technology means and how it integrates into daily life, what it means to be human in our modern world. Sometimes technology is seen as redeeming our lives: nondisabled people believe—and expect us to believe—that technology will "solve" the problem of our disability and save us, or those like us, in the future. Yet these expectations often don't match our circumstances. They confine us. When people assume that one device will "fix" us, they don't pay attention to the host of other concerns around disability technology—the bad planning and design, the need for constant ongoing maintenance, the problem of money (Jill's van, like a lot of disability tech, wasn't paid for by insurance), and the staggering lack of social support for disability accommodations (Jill's apartment managers, the police). These are all forms of ableism.

Talila A. Lewis describes ableism as "[a] system that places value on people's bodies and minds based on societally constructed ideas of normality, intelligence, excellence, desirability, and productivity."[4] Ableism is a system set in socially constructed norms. Ableism is more than just bias: it's the entire idea that anything can or should be perfect in this universe of entropy and chaos, applied at the level of human bodies and ways of being. Technoableism is a specific strain or type of ableism, this deeply embedded force for valuing limited ways of being. Like ableism writ large, it informs how we understand and process the lives and stories of disabled people. It informs how we decide who is worthy, who is entitled or deserving, and what justice means when it comes to technology and intervention. It reveals itself in how we rank and value disabled people against each other, and how these values shape our institutions and infrastructure.

In the constant praise and promise of bodies-fixed-through-technology, we see that disabled is a bad state of being, and that disabled people must be altered to be worthy. The Cyborg Jillian Weise writes:

> They like us best with bionic arms and legs. They like us deaf with hearing aids, though they prefer cochlear implants. It would be an affront to ask the hearing to learn sign language. Instead they wish for us to lose our language, abandon our culture and consider ourselves cured.

The sharper end of technoableism is that, if one cannot measure up by technological means, then there are technological-eugenical means for "dealing with" the most unfortunate cases. Prosthetic legs, Prozac, pacemakers, ostomy bags, fidget spinners, heating pads, and wheel-chairs are technologies for disability—and so are gas chambers, insti-tutional confinement, and prenatal de-selection.

I've been collecting my thoughts about technoableism, in one way or another, for the past decade, so my understanding of the phe-nomenon has been filled with constantly accreting examples and expe-riences. Most of the chapters in this book are stand-alone, intended to be readable on their own or in any order. (Why should people be forced to move linearly through the text? Crip aesthetics value nonlinearity.) Yet they all offer "ways in" to conversations about technology, disabil-ity, cyborg life, ideas about minds and bodies, and classic themes in the philosophy of technology as it relates to disability. For those not already engaged with these areas of interest, I hope this book will be a useful springboard into various topics. While I draw from fellow scholars, I also draw from blogs, fun with friends, Twitter, entertainment, media, and my own personal experiences as a technologized disabled person—a cyborg, a cripborg.

Throughout, I consider disabled people the experts about disability. I'm no longer interested in what so-called experts (nondisabled scien-tists, physicians, therapists, and so on) have to say. These narratives are already overrepresented, and in some cases, they have done damage to disabled people as a community, disabled people as knowers, and the experience of disability as valid and valuable. When we don't listen to those with actual experience, we often get accounts of disability *and* technology completely wrong. Disabled people are "the real experts" (the title of Michelle Sutton's 2015 edited volume about and authored by autistic people) when it comes to technology and disability. We use technologies. We also reject them, grapple with them, or repurpose them. The views on technology we get from listening to disabled people often look very different from those of people educated in the medical and "helping" professions.

Likewise, the language we choose for ourselves differs from the language others apply to us. Think of how often disabilities and procedures are named after the physician who "discovered" a condition or technique rather than the people who experience it or those who were the subjects of "pioneering" experiments into disability. We are named and claimed by experts who may know our bodies in limited ways, but rarely know our communities, our experiences outside their clinics, or our reception in the world.

The story I want to offer is not clean, smooth, or linear. It's bumpy like tactile pavers under my walker, shifting like the gravel that rolls and slides under my prosthetic foot, slippery like unplowed sidewalks in the winter. Part of this is informed by my own neurology, as someone with chemobrain. In the "Gotta Catch 'Em All" spirit of Pokémon (we will address Pokémon as a disability technology in Chapter 5, by the way), I am a hard-of-hearing, chemobrained amputee with Crohn's disease and tinnitus and possibly some undiagnosed anxiety or PTSD related to my cancer treatment. Few disabled people have just one disability. That's important to note because most studies about disability want you to have just one—which means most studies about disability do not record our data with any fidelity. The word "chemobrain" was coined by former chemotherapy patients, and the neurological change was first recognized by patients, decades before it was "validated" or believed by medical experts. Chemobrain refers to the effects of systemic chemotherapy on the brain—sometimes short-term, sometimes permanent. It involves changes to memory, focus, information processing, and cognition. I am grateful to ADHDers and autistic people, who developed and promote the neurodiversity paradigm, which includes under its umbrella all kinds of people with non-normative neurologies like mine (see Chapter 5) and helps me be okay with myself.

There is so much we learn in the disability community from each other—not just in our own narrow disability communities, but in cross-disability conversation. I hope to share some of what I have learned. Not just because knowing is good but because I want there to be more space for disabled existence, disabled creativity, disabled planning, disabled flourishing, disabled life, disabled expertise, disabled everything.

A QUICK GUIDE TO THE UPCOMING CHAPTERS

Chapter 2: Disorientation. This is a good start for those unfamiliar with disability studies and disability activism. I hope to "dis"-orient you—and not just through clever puns. I want to shake up the tropey narratives this chapter's touched on already, and reorient readers to disability-forward, disability-led conversations about disability, disability history, and disabled community. So much of what people think they know about disability comes through the lens of nondisabled expert-scientists about disabled people, and that's not good enough.

Chapter 3: Scripts and Crips. In this chapter, I really dig into those tired stories about disability, writing about the common "tropes" or archetypes that appear in stories about disability in news and entertainment media, social media, and memes. Identifying these tropes can steer you clear of a lot of cringe and help people do better by disabled people.

Chapter 4: New Legs, Old Tricks. Here I stay with mobility disability for a bit—the disability that I have the most experience with, and the one that first brought me into community with other disabled folks. This chapter is built around anecdotes from wheelchair users, arm amputees, and leg amputees to counter the unending media narratives that focus on physically disabled people. We see constant coverage of Paralympics and TED talks, and other public narratives of disability usually focus on young, white wheelchair users or amputees. Here I write about the pressure to be (or appear) normal—to walk, to have a five-fingered hand—and explain the real process of getting a new prosthetic limb.

Chapter 5: The Neurodivergent Resistance. I define some key terms about neurodiversity and the neurodiversity movement, then talk about the eugenic history that led us to our current landscape of disability charity work, disability techniques, and disability technology. I address one of the most common approaches to autism intervention and turn to look at what autistic people recommend as autistic-led, autistic-friendly autism approaches and preferred technologies. As someone who is not autistic, I draw heavily here from autistic blogs, books, panels, and friends.

Chapter 6: Accessible Futures. My final chapter turns to ideas about outer space, bodies, disability, and technology. I argue that disabled existence is expansive; it provides for creative possibility, especially when planning for more habitable futures for us all.

CHAPTER 2

Disorientation

I LOVE AMPUTEE COALITION OF AMERICA meetings, less for the educational sessions than for the experience of just being in a place where everyone is like me in some way. As I walk around the conference hotel, usually with my friend Mallory Kay Nelson (we road-trip it together, and I have not been to a convention without her), we end up casually looking at everyone we pass to see which limbs they are missing. It's always a bit of a surprise to see non-amputees. Here, the nondisabled people are the minority—the odd ones out, not us. It's honestly a relief.

At the conference, we sometimes get silly and try on each other's prosthetic sockets—although they never fit anyone else but the person they were molded to. There are sessions for amputee swimming and rock climbing and running, and nice as these activities are, the real highlight is the other people. I love the convention floor. There are people who want to sell you their expensive and snazzy feet; they let you see them and play with them, and after the event you can ask your prosthetist about that brand and type. The feet booths are usually pretty busy—most amputees are leg amputees, and nearly all amputees with leg amputations (whether below the knee, above it, or at the hip) will need prosthetic feet if they choose to wear prosthetics. There's a booth that displays specialized underwear to relieve prosthetic rubbing in delicate areas, some with padding for people who have lost butt cheeks or have a tendency toward pressure sores. I like to watch how people strategically avoid that booth. People do talk to the sales rep, but they tend to take the literature and go rather than play around with the display goods the way we do with feet.

There's an amazing table of arm doodads with demonstrations by an experienced arm amputee. She shows people how to knit in the round, what kinds of cutting boards can hold your fruit so you can slice it without it rolling away, tools for using buttons, and some other life hacks. Hers is a noncommercial booth that does a community job. She's not selling anything; she's just helping newbies (and nosy folks with leg amputations like me who want to know more about arm-amputee life) see the array of tools that can make tasks easier.

The Amputee Coalition isn't perfect; nearly all of us still have some internalized ableism, and there's a clear hierarchy in some conversations. There are some people there who buy into the "inspiring disabled person" narratives; these folks try to hold themselves up as examples and inspiration rather than participating in community in richer ways. And there's a bit of a divide between people who aren't interested in wearing prostheses and those who are. For some people, especially people with high-leg amputations, some arm amputees, and people with other damage to intact limbs, prostheses are more time and effort and expense than they are worth. Some of these folks have already sunk a lot of time into the process of trying out prosthetic legs or arms, only to discard them because of the extra weight, medical risk, or plain old discomfort.

My friend Mallory tried and used a prosthetic leg for a while, but as a hemipelvectomy amputee (someone whose amputation starts at mid-hip), there's not much to hang a prosthetic leg on. The parts for hip-disartic and hemipelvectomy prosthetics are heavy, requiring three different prosthetic joints: hip, knee, and foot. Hips and knees are wildly expensive, and even the most common and reasonably priced part of the setup—a below-the-knee foot and shin—isn't cheap. My last leg was $13,000 (billed to insurance). Mallory's entire setup would run in excess of $100,000. Because her amputation is so high, she also had to buckle her prosthesis around her waist and deal with a lot of rubbing and discomfort—more than someone with a lower-level amputation can fathom. The fitting process can be terrible in such sensitive territory. (We are also in vulnerable positions when we seek some technologies. Another friend of mine, an above-the-knee amputee, had a prosthetist who groped her. She sought prosthetic work elsewhere and filed reports with the police, the Better Business Bureau, and prosthetics certification groups, but the man is still in business, unscathed.)

Instead of prosthetics, Mallory uses both a wheelchair with a specialized cushion (to support her half-assed ass and not mess up her spine in the process—a normal chair would not work for her) and a pair of forearm crutches (which are better than underarm crutches). She calls herself *transmobile*: in her universe, there's no one perfect technology, so she has a whole array of technologies she can use in different situations and to fit different needs. She has *more choice* than most people about how she moves in the world, and I love how she has shifted my perspective on this. She has shown me that being disabled gives us more choice than nondisabled people have. It's a generative thought, and one rarely voiced outside our community.

It's very common for disabled people to write and talk about technology in ways that contradict popular media narratives about disability and technology. Mallory is one example: she uses a whole stable of technologies, depending on her current needs.[5] Yet the media often gives the idea that disabled people all seek one perfect replacement technology. Prosthetics are the most visible example of this "one technology that works for every situation," but we see this narrative even in discussions of congenital and developmental disabilities. Most of the stories we are told about disability and technology come from nondisabled storytellers through entertainment and news media.

It should go without saying that we need to center disabled people as experts about disability and technology. Yet if we do, we really trouble some underlying assumptions of the ableist world we're in. In its simplest definition, ableism is bias or discrimination against disabled people or stigma against the status of disability—a bias toward nondisabled lives and ways of being. Anti-disability bias often underlies or reinforces other types of biases. Saneism (bias based on mental traits), audism (bias against deafness), and fatphobia are all manifestations of ableism. Perhaps less intuitively, white supremacy, racism, sexism, and xenophobia are also intertwined with ableism. For example, ableism is bound up in the ugly history of scientific racism that was used to justify prejudice against non-white people. IQ testing and racial measurement norms, both constructs of nineteenth-century scientific racism, still underpin how people see disability today. Ableism has also historically been used as a justification for racism and xenophobia in U.S. immigration policies, which depicted some Western Europeans as superior in intelligence, healthier and more resistant to disease, better citizens less motivated to crime, and all-around better candidates for immigration than those from elsewhere.

This type of thinking still informs modern immigration policy. Most disabled people are not allowed to immigrate to become a citizen in other nations.[6] Disability automatically casts you as a burden, making you ineligible for permanent citizenship or residency in many countries. When nondisabled friends lament the current political condition in the United States and talk about moving somewhere with more progressive social policies, they don't realize that this is a privilege that most disabled people can't even dream about.

Ableism is deeply entrenched in language, which reflects how we think about things. When we use everyday insults like "idiot" or "moron," both of which were originally medical diagnoses of mental disability but now are used as pejoratives, we're employing language that reveals what we think about disability. Similarly, when metaphors of blindness are used to mean ignorance or other negative attributes (despite all the competent and knowledgeable blind people who exist in our world), it encodes the assumption that disability is limiting. Blind linguist Sheri Wells-Jensen explored many different languages, trying to find some use of the term (outside its use to describe literal physical blindness) that didn't have a negative meaning. She found that this linguistic stereotype—blindness as ignorance—is nearly universal across languages.

Ableism is also written into the law. Many laws encode ableist discrimination, such as legal provisions that allow kids to be taken away from disabled parents by child services more easily than from nondisabled parents, for no other reason than the parents' disability. Historically, municipal laws known as "Ugly Laws" criminalized disabled people for simply existing in public spaces—for being visibly disabled—as if our appearances were inherently uncivil. These anti-disabled laws very much coincided with racism and alarm about vagrancy, giving authorities an excuse to pester people they did not like the look of and arrest them for being in a public space at all. In the United States, the last of these laws was not repealed until 1976.

I tell my Technology & Disability students frankly that my class is biased, but in the opposite way from most of their other training. Many of my students are in training for the health professions, engineering, or science. Virginia Tech's slogan used to be "Invent the Future,"[7] and they believe they *are* part of making the future. I love it because I believe it too. My job as a humanities educator is to make sure they have the right context and understanding to do their work in an ethical, socially beneficial way. Too many people going into these professions want to go and design and make and therapeutize and "help" without ever having a deep understanding of the people they want to work with and for—without understanding the context of our lives.

In my class, disabled people are the experts about disabled people. We don't read nondisabled accounts of disabled life, because these accounts have so often gotten our stories wrong. When we think about autism, we don't consult the *Diagnostic and Statistical Manual* (DSM) or ask autistic people's families about autism: we listen to autistic people themselves. Autistic adults are out there, many of them. They've written books like *The Real Experts*, edited by Michelle Sutton, and *All the Weight of Our Dreams: On Living Racialized Autism*, edited by Lydia X. Z. Brown, E. Ashkenazy, and Morénike Giwa Onaiwu (see Chapter 4). Organizations led by autistic folks like the Autistic Self-Advocacy Network (ASAN) and the Autistic Women and Nonbinary Network (AWN) do amazing work that truly values autistic people.

By listening to disabled people, we learn a whole new way of thinking about disability. In contrast, many charitable organizations (most of them run by nondisabled people) like Autism Speaks promote what is known as the "medical model of disability," which often impedes flourishing of disabled communities. (I will talk more about this in Chapter 5, but Autism Speaks is highlighted as an anti-autistic group by many autistic activists and self-advocates, who have spoken out for many years about the way this organization is managed, its philosophy, and its lack of respect for autistic people.) The medical model is the idea that disability is a malady, something *outside the norm* that needs to be addressed, cured, eliminated, or remediated through medical or therapeutic intervention. Under this model, disability is something to root out, something to work against, something to fear or pity. Disability is framed as a *problem* that resides in individual disabled people—a problem that needs to be addressed. The solutions to this problem are different in different stories: examples range from early therapy interventions to technology like a prosthesis to preimplantation genetic diagnostics and elimination. Under this worldview, the role of a professional who "works with" disabled people is to identify the proper therapy, technology, or approach and deploy it in order to solve the problem. Nondisabled experts about disability make decisions about what disabled people need, about how to assess us and on what criteria, about where we belong (and make no mistake, they can institutionalize us if that's where they decide we should be); about what benefits and technologies we qualify for; and about the trajectory of our lives—our life prospects, our job prospects (some of us are often pipelined into specific types of work); our education (including whether it's worth giving us any education at all); even whether we can marry or have children. The Supreme Court case *Buck v. Bell* (1927) upheld the state's right to forcibly sterilize disabled citizens, and it has not been overturned.

There are lots of problems with the medical model. Foremost, of course, is that it doesn't let disabled people talk back or have agency or make our own meaning. But there are other, more philosophical problems. For example, how do we even create a coherent category of disability? How do we define what counts as a disability and what doesn't? The category of "disability" doesn't fall simply along the lines of impairment: we don't count most people who wear glasses as disabled, although they are visually impaired. And not everyone who is disabled is impaired: some people with dwarfism who are otherwise healthy experience the world as disabling simply because things are built too tall—from shelving in libraries and groceries to airport and bank counters to high-top restaurant dining (which, to be fair, is awful for most people). Categories of disability are constructed relative to our expectations and norms.

As a good dis-orientation tells you, disability is a social construct—a mismatch between the self and a world that was designed to cater to normative bodies and minds. Disability is a made-up category. Of course, this made-up category has real effects, with enormous social, cultural, and personal significance and consequence. I'm obviously not saying disabilities aren't real! I'm an amputee, and long before the idea of disability as a social construct captured me and others, people were born without limbs or were losing limbs—amputees show up in the fossil record. But why are an amputee, a person with dyslexia, a blind person, and someone who is bipolar all in the same category? (This sounds like the start of a joke.) What holds this category together? Disability, which today is a category of understanding, is actually a historical concept that developed relative to work, employment, and education. Historical and social factors underpin how disability is defined and how people are grouped.

The wheelchair is the universal icon of disability. It's on restroom doors, parking spaces, and ramps. Yet technologists are always trying to replace the wheelchair (which is itself a piece of technology). Exoskeletons and devices aimed at walking and climbing stairs—designed to make disabled people adjust to the world as it is, obviating the need for ramps or elevators or accessible doorways—are by far the most commonly covered mobility devices in mainstream media. The wheelchair, instead, requires the world to adjust to the disabled person.

The Social Model of Disability insists that disability is a *social* phenomenon: the problems are not in the bodies or minds of people but in the stigmas and barriers erected by society. According to this model, people count as disabled or abled depending on social context, social structures, and the built environment. The categories of disabled and not disabled depend upon notions of normalcy, as well as philosophical ideas of what it means to be human and who is deserving of rights. Even today, definitions of disability can be porous. Some amputees deny that they "count" as disabled after they master prosthetic use; some Down syndrome advocates remind the public that not everyone in their community has extensive medical problems; and some little people point out that their biggest barriers are those that come from social stigma and those created by the built environment. These categories vary across cultures and over time. In fact, there are cultures and historical periods that have had no category of "disabled"—mostly times when many *more* people had what we would now call disabilities, due to disease, frostbite, or other harsh conditions of life. Today, many of our ideas about able-bodiedness and disability come from classifications based on who is suitable for plantation or factory work: we call people "disabled" when they can't perform "normal" amounts of physical labor.

I am not saying that disabilities don't cause problems for people—

this isn't the point of the social model. Many conditions associated with disability do cause difficulties, and disabled folks willingly participate in rehab and physical therapy and other therapies. The social model simply asks us to expand our ideas about disability: it pushes back against the knee-jerk assumption that disability is abnormal and that our bodies and minds should be normalized. In fact, disability is extremely normal in the statistical sense of the term: more than 15 percent of the current population has some form of identified disability—psychiatric, learning, developmental, cognitive, sensory, or physical—and nearly everyone who lives long enough will join this group; aging itself produces disability, as it is currently defined in our culture. This is why some disability advocates used to call nondisabled people "temporarily able-bodied—or TABs."

Disability has not always been as deeply tied to the ability to do labor as it is now. As Kim Nielsen's *A Disability History of the United States* shows, many Native American cultures had more room for variation and a different sense of health and wellness;[8] disability was (and is)[9] not stigmatized in the same ways among some Native American groups (and, incidentally, many Native languages have no word for "disability" as a category). Nielsen explains how Plains Indian Sign Language (the most well documented of Native sign languages), for instance, was used by various Plains tribes as a common language of trade—so signed language was a natural part of culture and language such that communication norms allowed more easily and naturally for deaf inclusion. Marisa Leib-Neri explains Native American concepts of wellness and health, writing:

> Native American beliefs are fundamentally rooted in ideas of a multi-
> layered and ever changing reality. Because reality is ever changing

Native Americans have no conception of normal and conversely have no concept of abnormal. It is important to keep in mind that like the larger American population, Native Americans and their respective tribal nations are not entirely homogeneous. However, historians have been able to draw extremely similar conceptions of disability, health, and wellness among tribes. The common thread of many Native American tribes is their ability to embrace difference in a variety of ways.

Colonization deeply shaped understandings of work and community, shifting ideas about disability by introducing new stigmas. For example, Indigenous North Americans—whose customs, traditions, values, and even norms of communication differed from the dominant European colonial culture—were often read as psychologically unfit to be self-determining. This invention of disability had horrific material consequences for Indigenous people in North America; we are just now starting to reckon with some of these horrors as we uncover mass graves of Native children on former Indian Residential School sites. These schools were intended to "reform" Native children to be part of "normal" (read: white, proper) society. Our governments, often with aid from Christian "charities" and mission work, sought to cure the "problem" of differing cultures by taking children away from their families and communities.

In the United States, the introduction of chattel slavery firmed up disability as a physical category, one centered on which bodies and minds have or generate economic value under taxing working conditions. Disability began to be mapped as the group of physical and mental conditions that limited the price at which enslaved people could be sold at auction. Some disabilities rendered people who could not do physical work, or bear children to work, as entirely without value. The historical record contains horrifying accounts of disabled Black and Indigenous people who were murdered because they were judged economically worthless. Other new mental disabilities targeting unfitness to work were invented: in 1861, a mental illness called drapetomania was used to diagnose enslaved people who tried to run away and escape to freedom as mentally ill. Like Native people, Black people who refused to capitulate to colonizer culture in order to survive were categorized as mentally disabled. The category of disability as a fundamentally economic judgment continued to be refined through the era of industrialization, when certain types of people—those who were not up for long hours standing at machines or doing small, repetitive tasks—were deemed disabled.

It's hard to overestimate the ways in which racism and ableism have shaped where we are today. The recent disability justice movement has sought to trouble the economic categorization of disability, reimagining it as a category of political solidarity. The movement is banding together different groups of disabled people to rally for political action, recognition, and change, claiming disability as a political identity and cause. By embracing disability (rather than distancing ourselves from it, as many of us are encouraged to do), the movement has built coalitions to work toward a more equitable and inclusive world.

WHEN WE CONSIDER HOW we socially, as a group, frame the world that disabled people must navigate, we also begin to think about the language we use. There are contentious debates among disability professionals, disabled people, and different disability communities about how to talk about disability, particularly what we should call *groups* of people (it's fairly easy to ask individuals what they want to be called). Is person-first language or identity-first language better? Do we say "person with a disability" or "disabled person"? As an amputee, I'm often "meh" about these debates. In the amputee community, the "person" part is basically left implied: "person with an amputation" sounds like I need immediate medical attention to reattach something, "amputated person" sounds like they cut me off of someone else's body, and "amputee person" just sounds terrible. "Amputee" is fine. It's important to call people what they want to be called. However, there is a lot of history around terms for disability that nondisabled people who want to be sensitive should note.

Historically, disabled people have been described using terms like "unfit," "invalid," "crippled," and "handicapped." These all denote slightly different things. Unfit is more likely to contain congenital, developmental, and genetic-related disabilities; invalid is more likely to refer to chronically or physically disabled people and war-wounded; crippled refers most often to disabilities that affect mobility; and handicapped is a catchall for those disadvantaged by disability, in some cases including both mental and physical disability. Originally, many laws referring to disabled people used the word "handicapped." Now many have simply been updated with the new terms, essentially using find-and-replace to swap "handicap" with "disability." This includes laws still in effect today, like the Individuals with Disability Education Act (IDEA), which was known as the Education for All Handicapped Children Act from 1975 to 1990.

The word "handicapped" still circulates, mostly among older people who grew up with this being the preferred term, but it went out of fashion for at least two reasons. One thing that sank the term "handicapped" was the offensive etymology often given for the term (one I sometimes still hear people repeat as true). According to this false history, the term "handicapped" originated with beggars putting their caps out to ask for money: hand-and-capped, to ask for handouts. The term "handicapped" actually comes from sports but plays so neatly into stereotypes of disability that people who were challenging prejudicial laws and educational policy preferred not to use the term.

Therefore, disability rights activists in the 1970s chose the term "disabled" instead of "handicapped." They gave new impetus to what they called the disability rights movement. Personally, I love the word "disability" chosen by our leaders and heroes. As disabled author and arts consultant Simi Linton says, "In the beginning [after becoming disabled], I didn't know what to call myself. I think the operative word was 'the handicapped': it was a term I never wanted applied to me. So I started saying I'm a disabled woman, putting it front, disability front. It's up there. It's out there."[10] And the language of the disability rights movement leaders also shaped legislation. Legislative language changes based on who is in the room to discuss and promote and lobby for the legislation: while the Rehabilitation Act of 1973 used "handicapped," the 1990 Americans with Disabilities Act used "disabled." This has brought the term into wide circulation and shaped the public consciousness.

Lawrence Carter-Long has written beautifully about the etymology of the word "disabled," pointing out the deep differences between the prefixes dis- and un-. As he points out, *dis*-abled is different from *un*-able. He talks how one meaning of dis- is "to twain." In a popular Facebook note from 2015, he writes:

Twain literally means "in different directions, apart, asunder." Use of the prefix in this way has given us perfectly good words like "discern," "discuss," "dismiss," "dissent" and "distill." When used in this way being disabled does not suggest a lack of anything—including ability, except perhaps to the uninformed or willfully ignorant.

Instead, it simply means diverging abilities—*differences*. (In the terms of philosopher Elizabeth Barnes, *disability* describes a mere-difference, rather than a bad-difference. An emphasis here on *dis*-ability not being *in*-ability exists too.) It is the prefix that makes so many people uncomfortable, for they seem fully willing to talk about "differing abilities," "handicapabilities," or other euphemisms that do away with the dis-. Lawrence Carter-Long's etymology here is part of his activism around language: he is part of a group of disability advocates who work on the #SayTheWord campaign, which urges people to stop using euphemisms for disability. As two-time Paralympian and scholar Anjali Forber-Pratt explains, "This movement encourages everyone to use the word 'disability.'"

I love that the word came from our community. Reclaiming the word "disability" as a part of disabled identity is at the center of the contemporary disability rights movement. It is also meant to educate others about how made-up words or euphemisms about disability are harmful. "Disability" invokes a political community in a way that "handicapped" does not. Bioethicist Rosemarie Garland-Thomson writes for the *New York Times* (where she has co-curated a series of opinion pieces by disabled writers) about her own transformation to seeing disability as a category of identity: "What has been transformed is not my body, but my consciousness." As the disability community notes, euphemisms like "differing abilities" elide the realities of disability (some of us can't do some things in some ways, and that's okay) and work against disability identity and pride (it's hard to have a pride movement when people can't even say they are disabled). There are also practical reasons to use the word "disability" rather than a euphemism: the shared term allows us to find each other and find what we need. It is important to be able to google for communities, medical needs, therapies, and more. Using euphemisms makes things a lot harder to find!

Working against paternalistic and euphemistic restylings that talk around having to say "disability," the #SayTheWord campaign is meant to unite two groups of people in the disability community: those who prefer person-first language and those who prefer identity-first language. Person-first language syntactically puts the person before the disability, quite literally: phrases like "person with a disability" rather than "disabled person." This is the preference of some disability communities, like those of people with intellectual disabilities. Identity-first language—"disabled person"—is the preference of many other communities and used by Deaf organizations, blind organizations, and communities of autistic adults, among others. #SayTheWord calls a truce on this debate. It does not dictate to any community or individual what they should call themselves, simply asserting the importance of using specific, clear terms such as "disabled": it insists that "disabled" is either a value-neutral term, or one of pride and identity, not a terrible fate or label about which one should feel ashamed.

As an amputee, I vastly prefer the identity-first construction for myself. "Person with an amputation" makes it sound like I'm about to bleed out—but don't call an ambulance, I'm already an amputee! It is also linguistically awkward to list my multiple disabilities—"I'm a disabled person" covers it all. It takes too long to say "I'm a person with an amputation and Crohn's and tinnitus who is hard of hearing and has chemobrain." Indeed, most of us don't have just one disability. My cancer treatment led to three of my disabilities, and some of us end up at higher risk for other conditions (for example, autistic people are more likely to have digestive issues); the same traumatic event can cause both physical and mental disabilities; and being disabled can put you at higher risk for other conditions due to chronic ableism and disregard.

There is another term for disability that some disabled people like: "crip"—a derivation from "cripple." "Cripple" is a derogatory term for a disabled person. However, there's been a movement to "reclaim" the word "crip," in much the same way as we see the LGBTQIA2S+ community reclaiming the word "queer." We see it in the names of a lot of blogs, like Bill Peace's The Bad Cripple, Karin Hitselburger's Claiming Crip, Ingrid Tischer's Tales from the Crip, and so on, as well as in the 2020 film *Crip Camp*, which was co-produced by the Obamas. Like "queer," it has also been verbed, and is sometimes used to describe the process of disorienting and undoing ableist ideas and structures. We also see "krip" with a *k* in the Krip Hop Nation, a collective of black disabled artists that includes such luminaries as Keith Jones and Leroy Moore Jr. Moore has used the term "Krip Hop" since the early 2000s.

"Crip" is also used by disabled political activists Alice Wong, Andrew Pulrang, and Gregg Beratan to describe their series of articles on and campaign for disabled voting and disabled political issues, which they titled #CripTheVote. This is a nonpartisan effort to turn out the disabled vote and educate and sway political candidates on issues important to disabled people. They are direct about why they chose this hashtag:[11] "Selective use of 'crip' or 'crippled' by people with disabilities is a conscious act of empowerment through 'reclaiming' a former slur as a badge of pride." Like the term "queer" two decades ago, this might not yet be a term everyone can use, but it is appropriate for disabled people describing themselves, their communities, and how they do things in the world—"cripping" it.

The bottom line on language is that disabled people are a huge and varied group, and we're not all going to agree. That's good, healthy, and appropriate. But the changing language around disability reflects the recent flowering of disability arts, culture, and critique that are explicitly disabled-driven (cripped) and were made to convey disability pride and power. I think of the exciting exhibition by the Paul Longmore Institute at UC Berkeley on disability history, called *Patient No More*, which speaks so strongly to disabled people being more than patients (and indeed not always patients!), more than objects of medical interest. The disability rights movement and the social model of disability overturned the patient-only model of disabled people that saw us as the domain of medicine—as the passive subjects of diagnosis and treatment to solve the "problem" of our disabilities. Now we have agency: we talk about what we want—our fit in societies, our lives beyond clinics, our flourishing in the world, our community and culture. We can use language to indicate defiance in the face of ableist norms and pride in community, history, and solidarity.

In the novel *Accidents of Nature* by Harriet McBryde Johnson, there's a scene where disabled young women Jean and Sara are talking together at their disabled summer camp. Jean is waiting for assistance for an after-swim shower; Sara hadn't gotten in the lake and doesn't need a shower, so she sits on her cot, paging through Jean's scrapbook from home while Jean waits. The scrapbook has newspaper clippings from Jean's local paper with stories about her. Sara reads to a third girl, Margie: "Crosstown Cripple to Lead Fire Parade . . . makes Honor Roll . . . gets award . . . Poster Child" (the scrapbook includes flyers for telethons for which Jean had been the literal poster child). There is also a clipping with the headline: "Crosstown Girl Learns to Walk." The story includes Jean's years of physical therapy, operations, and leg braces, and Sara reads the whole thing aloud. She then says to Jean:

It's funny. Therapists, teachers, relatives—everyone—they all think walking is such a wonderful thing. And we don't question that. We believe it must be worthwhile or they wouldn't torture us for it. And then, finally, you get up on your feet, take a few halting steps—pardon me, I mean courageous and determined steps—and the cameras flash, and everyone is inspired. But then you find out walking is a lousy way to move from place to place. And as you get bigger, it's worse. When you fall down, you have further to go. When you start to think for yourself, you realize a wheelchair is a better way to get where you are going.

Margie says wheelchairs are nice and plays with Sara's wheels; Sara describes her own experience in physical therapy, expressing relief that her legs were too weak to force her into much of the effort around walking.

It's incredibly rare to find narratives like this one. In media and in wider conversation, we almost never see depictions of disability where normalization is not a goal—that is, where people are not made to use technology to walk or stand and are glad for it. Sit with this idea for a moment. So many of our stories about technology and disability are about technologies as redemptive, as having the power to normalize disabled people, to make us "overcome" our disabilities. They show us "better" living through technology, where better means something pretty specific in how people exist in the world. But sometimes technologies don't make us better—indeed, the perception of *better* here is based on ableist norms, not on the desires of disabled people. Sometimes the "fix" ends up being too painful or takes too much time; sometimes it is fleeting. That moment of glory, captured in Jean's headline of

"Crosstown Girl Learns to Walk," was temporary, for Jean is a wheelchair user when she is at camp with Sara and Margie.

Jean begins to see the problems with this attitude—that disabled people must always strive to be as "normal" as possible, to learn to walk—when she thinks to herself, "My parents saved that article about me to celebrate a brief triumph; all these years it has remained in my scrapbook as a permanent reminder of my only real failure, a disappointment none of us wanted to talk about."

THE WAY WE TALK about and portray disability and technology encodes so many expectations about how disabled people should think about therapy, technology, and life. There's an expectation that disabled bodies and minds (bodyminds, if you are a scholar of disability studies) call for amelioration, for fixing, for specific types of care. But these imagined calls rarely center disabled people's experiences and desires. We rarely recognize disabled expertise about what works and what doesn't, and what good approaches to tech and disability look like for disabled people. Most disabled people certainly don't hate technology, but many of us resent the tropey-sappy depictions of technology that center abled saviors and their impressions of what we need.

CHAPTER 3

Scripts and Crips

■   ■   ■   ■   ■   ■   ■

PARALYMPIAN AND POET LYNN MANNING, a blind Black man, has a wonderfully illustrative poem about stereotypes of disability, called "The Magic Wand." In this poem, Manning describes the way people around him think about him from before and after he unfolds a cane that reveals to others that he is blind. As a tall Black man, he already bears the weight of biased expectations about his character and life from those around him. People expect him to be very good at basketball, or to be a scary criminal, a pimp, or someone who has gotten rich off the public dollar—the ridiculous stereotype of "welfare queens" from the 1990s. (Quick sidenote: It's really hard to get "on disability"—as in, receive Social Security disability or some other form of disability welfare. Disabled people are required to offer a lot of proof about whether they are truly disabled and whether they truly deserve the support. A lot of people need support who currently don't get it! In addition, disability is an absolute pittance, almost impossible to live off of and can take many years to apply for and appeal denials.) But then Manning unfolds his white cane. White canes are what many blind people use to navigate, but they also signal their disability to others. If the cane has red stripes, it indicates a person is deafblind. Once he unfolds that cane, people see him not as a Black man but as a *blind* Black man, with a whole new set of stereotypes and expectations: being especially musical and good at the piano, being especially wise, and being pitied as a child who no one wanted, a burden to those around him. It's not that no blind Black men have ever been great at piano—this stereotype exists in part because Ray Charles and Stevie Wonder are nearly the only representations of blind Black men we have in our media sphere.

There's something to be said here about the lack of varied representation of disabled people, and especially about our invisibility when we don't fit particular stereotypes or narratives. There's also something to be said about how stereotypes about disability are entangled with ideas about race, gender, and sexuality. When disability is represented at all, nearly all the representations are of white disabled people. Dom Evans and Ashtyn Law have been working on a project called FilmDis (filmdis.org). They have tracked representations on TV of disabled people over various calendar years and written reports about disability representation. As they point out, even when disabled characters are represented onscreen, they are usually written by nondisabled writers and played by nondisabled actors. So, when we see disability in entertainment at all, stereotypes tend to stick around.

These stereotypes are often expressed through "tropes"—repeated themes or motifs that appear in media (including not only entertainment and news media but also in science, engineering, medicine, education, and more). Tropes are talked about in literary, film, and TV criticism. (You may have heard of tropes like "bury your gays" or "enemies to lovers"—they are narrative tools that structure stories and set up our expectations for what is going to happen.)

Some tropes are weirdly specific. When I work with students researching disability narratives, we find five main tropes as we sort materials about disability: pitiable freaks, moochers and fakers, bitter cripples, shameful sinners, and inspirational overcomers. These familiar, oft-repeated narrative arcs about the lives of disabled people show up in all sorts of media—books, film, viral social media memes, news. These tropes are stereotypes—limited and inaccurate frames of reference for people to consider the experiences of disabled people. They also get in the way of people listening to what disabled people have to say when it differs from these narratives.

Let's look at each of these tropes in turn.

PITIABLE FREAKS

The pitiable-freaks trope appears across all kinds of media. It's tied to the history of freak shows and medical display more generally; perhaps less obviously (but more perniciously), it underpins the marketing of many charity campaigns. In the pitiable freaks narratives, disabled people are cast either as objects of curious medical interest or as objects of pity and charity. Every prom season in the United States, there are news articles that feature the kindness and heroism of high school students who take disabled classmates to prom. There's nothing wrong with dating or dancing with disabled people, but that's not what these stories are about. These are stories—considered "feel-good" news—

where disabled people are cast as the objects of pity, upon whom others can act to feel good. But can you imagine being the person who makes the news because someone asked you out? Treating disabled people going out for a date or at a dance as news is built on dehumanizing assumptions about who is dateable or interested in romance.

This trope shows up all over charity campaigns. Take the infamous "I Am Autism" commercials (which were put out by Autism Speaks, a non-autistic-led group that many autistic adults and their allies criticize). This group's fundraisers first depict autism as a sinister force in the world—something that breaks up homes and families, causing terrible fates for autistic people and those around them. The "pitiable freaks" trope that this plays on suggests that we must pity autistic people and act on that pity to solve the "problem" of autism. Autism self-advocates argue that these ads and others like them create negative, damaging public understandings of autism, far outweighing any potential good the charity might do with the money they raise using these tropes.

This same trope has long been identified and critiqued in the older Muscular Dystrophy Association telethons, which featured comedian Jerry Lewis. Former poster children—literally what they were called and where that term comes from—were called Jerry's Kids. In the actual telethons, Jerry Lewis waxed on about the horrible fate awaiting children with muscular dystrophy. The pitiable-freaks trope that underpinned these telethons deeply affects how disabled people were regarded and understood, even influencing the information we have about how our own lives will play out: it was while watching one of these telethons that four-year-old Harriet McBryde Johnson found out she was expected to die young—thus her memoir's title, *Too Late to Die Young.*

MOOCHERS AND FAKERS

The moochers-and-fakers trope shows up in coverage about disability benefits or accommodations for disabled people, who are often portrayed as cheating and lying to get benefits that they don't "deserve." Almost every news story about someone using the Americans with Disabilities Act to sue for their right to access a service or space depicts disabled people as harassing a business or trying to bilk the system.

We see this same trope being expressed in meme form. For example, someone has used the Snarky Wonka meme template and superimposed the text "Bless Your Heart. . . Trying to Get Disability Benefits. I Don't Think Lazy Is a Disability." It's actually really hard to get disability benefits—deliberately so, precisely because of the assumption that people are looking to cheat the system. People seeking disability benefits and accommodation are often treated like moochers (just as people asking their doctors for medication for chronic pain are treated as drug-seekers). Most people who apply for disability benefits are denied and have to make multiple attempts. In this system, there is actually very little cheating, but the assumptions expressed in the trope deny people what they need to work, participate, or survive.

Because of this trope, disabled people are scrutinized for every small modification or accommodation. The question "Do you really need that?" is one that haunts people who use wheelchairs or canes part-time, either because of disabilities that have flares or because their diagnoses are unclear; many people simply find that a wheelchair or a cane helps them complete daily tasks. What does it matter if they *could* walk farther in the parking lot if it's going to wipe out the entire rest of their day or week? Leave them alone and let them use the darn wheelchair without cross-examining them. I have friends who won't use the disability parking that they need (even with the documentation and placard) because they are afraid of strangers yelling at them or attacking them because they don't "look" disabled. This trope keeps people from using technologies that would be useful. It makes physicians resist giving prescriptions for assistive devices that would be useful, lest the person become "dependent" on them. It means that disabled people must surmount incredible obstacles to gain documentation that they are disabled. It invites intrusions into one's medical privacy; for example, my university HR asks me for much more health information than anyone should have to hand to their employer. I could go on and on about the damage done by this ingrained trope.

BITTER CRIPPLES

The bitter cripple trope is highly visible in mainstream media. Think of characters like Lt. Dan in *Forrest Gump* and Captain Hook in *Peter Pan*. The "bitter cripple" character is so tropey that it can give away a plot twist: if a character has a hand amputation or a visible facial scar or difference, viewers immediately know that they are likely the surprise villain. Apparent disabilities are thus sometimes used to signify internal moral or characterological disability. Other times, the disability seems to cause the villainy instead of reflect it: it's also assumed that disabled people who are cranky are cranky *because* they are disabled. This trope keeps us from complaining about lack of access or inclusion, since we'll be dismissed as bitter; which lets nondisabled people off the hook from listening.

I've personally encountered this trope and seen it projected on others—where people are read as failing at "overcoming" disability by not having a good enough attitude (even within disability groups) or they are dismissed as overly combative because they are disabled. I'm lucky enough to work with other disabled people. Some of us are justifiably angry—like Bill Peace they are "bad cripples." I've been in meetings where the outspoken ones in our disability groups are not invited back after we didn't couch our feedback as calmly as others would like, or where what we say is ignored but top people applaud themselves for listening to our community. People often chalk disregarding disabled feedback and expertise up to disabled people being ungrateful or not properly civil. They criticize how we interact—after a lot of inequity and inaccessibility, sometimes with bluntness or frustration or anger— consciously or unconsciously drawing on the "bitter cripple" trope to ignore feedback and dismiss real community concerns. This trope also means that disabled people often have to moderate our justified tone and approach—to be good cripples—in order to be treated with fairness. Recognize that this is also true for many other marginalized people, and particularly Black women. Disabled Black women are at a particular crux when voicing concerns or giving feedback, or even just describing a situation.

SHAMEFUL SINNERS

Shameful sinners is a long-running trope that frames disability as a punishment or penance for some kind of sinful action—either the mother's or the disabled person's. This trope also emerges in the idea that a disability is a "challenge" intended to transform and humanize the people around the disabled person. This is especially visible in stories about the parents of disabled children, which almost always follow the "We thought having a disabled child would be a tragedy, but surprise! We found that we actually love our child! Aren't we angels on Earth?" narrative, in which the child is an occasion for the parents to redeem themselves morally. We see the first version of this trope play out in a lot of coverage of health and wellness, which blames people for their disabilities or conditions (She shouldn't have eaten sugar! He should have been a vegan! She should have exercised more!) and then denies proper access and care because they have a "lifestyle disease." (Disgusting phrase.) This shows up around lung cancer (if you get a lung-cancer diagnosis, people want to know, did you smoke?); around AIDS, which was once (and in some communities still is) depicted as a literal judgment from God for the "sin" of homosexuality or drug abuse; and now around long Covid (did you have preexisting conditions?). Similarly, the now-debunked "refrigerator mother theory" of autism blamed the mom for being too cold to her children and thus turning them autistic: autism was her sin played out in her child, her punishment for her emotional failures.

INSPIRATIONAL OVERCOMERS

The inspirational-overcomers trope might be the most common one in the media, and it deeply shapes our expectations for the normalization of disabled people's lives and abilities—all of which is bound up in our relationship to tech. After the first four tropes, all obviously malignant, the last one might seem less awful, but it's really not. Media grounded in the inspirational-overcomers trope is often called "inspiration porn" in the disability community: it can be seen in a million memes of amputees on running blades and other athletic-looking disabled people with captions like "The only disability in life is a bad attitude."

This trope implies that to be good disabled people, we have to be constantly fighting against our bodies or minds in some way. We must disown our collective or community identities in order to be individuals. Inspiration porn makes disabled people into objects to inspire others; we must make ourselves into spectacles whose purpose is to motivate nondisabled people or make them feel good.

Other disability tropes are used to motivate shame, create pity, justify policing, and more. The inspirational-overcomers trope (which usually features white disabled people) intensifies the effects of these other tropes, becoming part of our social scripts. Who is usually framed as a moocher and faker, and who is usually shown as inspirational overcomer? Stories of disabled white people (especially of the inspiration-porn variety) get a lot more airtime than stories of disabled people of color. As Sami Schalk and Vilissa Thompson have pointed out, white privilege pervades disability narratives—even the white disabled critiques of inspiration porn (some of which I make in this book) are made possible by white privilege. Thompson writes:

> I ardently despise inspiration porn, but as an advocate who focuses on the achievements and experiences of disabled Black girls and women (and other disabled girls and women of color), I do not have the "luxury" of picking and choosing stories that depict us in a positive, non-inspirational light. Why is that? Because our stories do not get the same or fair amount of spotlight and recognition as the ones that feature White disabled people.

Similarly, Schalk points out that analyses of inspiration porn should take into account the cultural norms and racial dynamics that shape how these stories are produced and received.

Inspiration porn usually shows disabled people, guided by humanitarian technologists and therapists, finding the right technologies to properly overcome the circumstances of our bodies or minds. This sets out a particular (often unreachable) goal for science and medicine and for individual disabled people. But why is this a bad thing? First, because the "overcoming" narrative pretends that structural problems of access can be overcome with the right plucky, can-do attitude on the part of individual disabled people; it lets abled people off the hook instead of calling on them to actually help make the world more accessible. As Stella Young puts it in her spectacular TED talk,

> "The only disability in life is a bad attitude," the reason that that's bullshit is because it's just not true, because of the social model of disability. No amount of smiling at a flight of stairs has ever made it turn into a ramp. Never. Smiling at a television screen isn't going to make closed captions appear for people who are deaf. No amount of standing in the middle of a bookshop and radiating a positive attitude is going to turn all those books into braille. It's just not going to happen.

The trope treats us as exceptional *individuals*, rather than as members of an underrepresented minority community that needs access and accommodations and structural change—not just gadgets to help us function individually. It exceptionalizes, rather than normalizes, disability as a facet of the human experience and exaggerates how different we are from nondisabled people. Not only is this dangerous, it's condescending and annoying—why are disabled people doing basic things lauded for merely existing in public? And it clearly works against inclusive design and disability rights. We shouldn't have to always be at our best—to put forth constant struggle and effort—in order to be allowed to participate. Instead, let's all (disabled and abled alike) think creatively about making adaptations so that everyone can participate in the world. Finally, the trope sets up our existence as being in service of other people (their empathy, their enlightenment, their "if they can do it so can I" motivation); it means that to be seen as worthy of care, respect, inclusion, and accommodation, we must properly perform a role: that of a "good disabled person."

I LIKE TO END MY first-day Technology & Disability class "disorientation session" with a passage from *Accidents of Nature*, the story of Sara and Jean at camp. Here, Jean is reflecting on Sara's decision to accept her chair rather than contort herself trying to walk, and wondering what it would have been like if she hadn't felt the intense pressure to walk like a "normal" person:

> Can a wheelchair be a choice, and not a failure? I'm not sure. Maybe Sara was abused by her physical therapist; surely for most people it makes sense to try to become as normal as possible. But what if normal isn't the only way to be? It's an oddly thrilling idea, but I'm not ready to put it into words.

Jean is taking the first step in the journey of rejecting society's normalizing expectations for disabled people—a road that is being paved by disabled scholars, activists, designers, critics, artists, and cultural workers who are putting these liberating ideas into words and action.

On the first day of class, my students get these messages: Invert the social pressures for disabled people to be "normal" or inspirational or worthy. Care what disabled people say, do, and think. Don't be neutral; the "neutral" medical position on so many things has justified our exclusion and incarceration, and the "neutral" social view makes us into unfortunate charity cases rather than fully human beings. Try being biased not against but *toward* disabled people in how we frame issues and problems.

CHAPTER 4

New Legs, Old Tricks

■   ■   ■   ■   ■   ■

FIVE MOMENTS IN BECOMING AN AMPUTEE:

FIVE—

After my diagnosis and subsequent start to chemo, a friend brings our family dinner. I tell her that I'm probably going to have my leg amputated, and she tells me to not think so negatively. But I can feel how big the tumor is if I run my hand along the sides of my thigh above my knee.

FOUR—

Everyone who is nondisabled has optimistic things to say about how advanced prosthetics are now. It's reflexive.

ME: "I'm going to get an amputation."

THEM: "Prosthetics are so advanced now!"

It's like a memorized church call and response: "Prosthetics are so advanced now" is the amputee-life version of "Amen." Friends share upbeat news stories about cutting-edge prosthetic technologies— including arm prostheses, which is interesting but not exactly relevant.

THREE—

I interview prosthetists on the phone in the weeks before my amp. I go with the one who has the most experience with rotationplasty, the type of amputation I will have. We have a long conversation. He tells me that walking will never be the same, and that I need to let go of any illusion that it will be. He's optimistic in ways others aren't: not knee-jerk "Prostheses will fix it!" but recognizing the limits of the technology, preparing me for the difference, tempering my expectations. I understand that my life won't be like what people see on TV or in the news. They don't show the parts I'm about to experience.

TWO—

I'm lucky that I didn't find my cancer by breaking my leg (which is how some people get their diagnoses). I dance with my kids as often as I can in the months before my amputation date, though for much of this time I am hospitalized hours away for chemo and am so worn out. My mouth is full of sores, and my middle sister, Betsy, has to come help prep at Christmas since I am so exhausted after a round of chemo. But I savor all the "last times" with my leg and my full mobility. In early January, perking up again, I walk quickly through the halls where I get inpatient chemo. A nurse in the hall comments on how speedy I am, and I say, as I keep walking past, "I'm getting an amputation, so I won't be soon." I don't stop or look back to see her face. I can't.

ONE—

I am home recovering from my amputation, a few weeks out. My youngest sister, Megan, is here with us for months, taking a semester off from college to take care of my sick ass—and keep things consistent and calm for my kids. I will have to get back to chemo, but while I am recovering from surgery I get a few weeks off from my chemo schedule, so I feel a little better every day. I am learning how to do old things in new ways. Megan spots for me as I work out getting myself from the house out to the car with my rolling walker (a "rollator"—this is a walker with wheels and a little platform to sit on; it also has a basket to carry things, a significant feature when you are using a mobility aid that keeps your hands and arms busy). I do one step with the rollator, get to the car, open the car door, and collapse the rollator while balancing on one foot and leaning on the car. I stick the rollator in the seat behind me and get myself into the front seat without help. It's a big day. I can get myself into my car all by myself. Megan still drives because I am on big drugs.

LIFT-OFF—

I have been interested in technology and disability for a long time, even before I knew I, too, would become an amputee.

The stories we see about amputation and paralysis are probably familiar to you, ingrained through Paralympic coverage and Toyota ads, through dance competition shows, through news articles about the latest and greatest in disability tech, through stories that feature war-injured veterans coming back to their homes and communities outfitted with our best tech. We are inundated with imagery of disabled people overcoming the circumstances of their bodies through technology, which is framed as a kind of technological salvation for bodies and minds. When I became an amputee, people kept reassuring me (actually, reassuring themselves) that with advanced and wonderful prosthetic technologies, I would be back even better than ever: super-human, enhanced, a ten-million-dollar bionic woman.

Other types of disabled people aren't always subject to this kind of narrative. They are supposed to wait in hope for the day when technology advances enough to help their "broken" bodies or minds, which are depicted as less worthy of love, care, pride, and positivity. In contrast, the amputee "problem"—the problem of being an amputee, particularly a leg amputee—is often depicted as being "solved." People point to Paralympian-turned-murderer Oscar Pistorius ("the Blade Runner") and Amy Purdy on *Dancing with the Stars*. The parameters for being a successful amputee (which generalize to parameters for being a successful disabled person, period) are clear: recover, overcome, inspire, become "normal" again but with the added razzle-dazzle of technological enhancement.

There are lots of breathless news stories about prosthetic technologies, exoskeletons, and technologies aimed at "fixing" disabilities. They always promise that the next big thing, the big breakthrough, will be *life-changing* for users—empowering them, allowing them to become part of society again, to function like normal people should, at play and at work. A few typical headlines:

"THE TECH GIVING PEOPLE POWER TO DEAL WITH
DISABILITY" —BBC NEWS

"HOW TECHNOLOGY WILL CHANGE THE LIVES OF
PEOPLE WITH DISABILITIES" —*FORBES*

"FIRST PROSTHETIC LIMB DESIGNED FOR WOMEN: 'I
FEEL LIBERATED'" —BBC2

"ROBOTIC EXOSKELETONS ARE HERE, AND THEY'RE
CHANGING LIVES" —POPSCI

On campus, I see my colleagues in engineering and computer science and "helping" professions teaching narratives about technology as miraculous, holding up to young engineers in training various prosthetic and exoskeleton projects as "life-changing" examples of humanitarian engineering, models of engagement with communities in need. Standing in the hall of one building on my campus, I see a wall of LCD TVs playing slideshows of the prosthetic hand prototypes designed by an Intro to Engineering class.

When I taught my Technology & Disability course in that building, my students stood and watched the changing slides, thinking about our readings from arm amputees about prosthetic devices and assessing whether the triumphantly displayed creations were actually useful for the people they were supposedly designed to serve. The bathroom-use hand to help lather soap seemed the most straightforward and potentially useful. As an assignment to teach engineering design processes, including prototyping and fabricating, this seems like a great exercise, and one with a limited enough scope to take on in such a class. I appreciate the creativity and hard work of colleagues in engineering. But I also see how media-driven coverage of such technologies has failed the intended recipients, those of us whose bodies are imagined, sometimes simulated, and used for interesting assignments.

Media coverage of "enhancement technologies" frames disability in pernicious ways. I'm tagged on Twitter by a fellow disability-tech critic to go check out an announcement: "New bionics center established at MIT with $24 million gift." One of the new co-leaders for the center, Hugh Herr (an amputee like me, but also not like me, since I don't program my own ankles or climb mountains) is quoted in an MIT News (read: public relations) article: "The world profoundly needs relief from the disabilities imposed by today's nonexistent or broken technologies. We must continually strive towards a technological future in which disability is no longer a common life experience. . . . I am thrilled that the Yang Center for Bionics will help to measurably improve the human experience for so many." Relief from disabilities *imposed* by technology? As he frames it, the problem is not that tech is created without drawing on disabled people's voices and experience but that the tech is "nonexistent or broken." His solution: not more thoughtful, well-informed tech but more, better, more advanced, more expensive tech.

I read this and think only of Bill Peace, the late bioethics scholar who blogged under the name "The Bad Cripple." He would have had the best response to this "news," but I don't have his wit. I'm just crabby. The thing is: many of us value our experiences as disabled people, value disability community—yes, even with the less glamorous aspects. We don't think our lives are inferior. Many of us don't think disability itself is a bad thing, so big talk about the world "profoundly" needing "relief" from disability "imposed by" technology seems askew. This is not to say that anyone desires a condition that is painful and/or limiting. But not all disabilities are the same, nor are they all painful or limiting or otherwise bad except in how they are viewed.

Rosemarie Garland-Thomson, known in academic circles for her "Case for Conserving Disability," provides some context for why I'm so cranky about the gushing PR for this new bionics center. As Garland-Thomson writes, "disability is inherent in the human condition," and it is important to our world. Disability is many things, not just visible differences or those that affect mobility, and we can expect more disability (not less, as Herr hopes) in the future. Even if the technologists' wildest space fantasies come true, everyone is disabled in space—with gravity changes, people lose familiar forms of mobility, as well as muscle mass, lung function, and mental clarity.

In more mundane futures, too, more people will be disabled. The environment is changing, and dramatic weather events are on the rise, which can cause chronic conditions like asthma. We will also see new diseases (such as Covid-19) emerge. Long Covid is an object lesson in becoming disabled through common illness. Post-viral syndromes— such as post-polio syndrome, shingles after chicken pox, the long-term health consequences of the 1918 Spanish Flu, Epstein-Barr's priming of some people's immune systems for multiple sclerosis—will continue to emerge and proliferate as new viruses appear, causing new clusters of symptoms and new types of disability. Eventually, everyone becomes disabled if they live long enough; like death and taxes, aging is inevitable. We should always be planning with disability in mind, because disability is an inherent part of having squishy meat bodies.

Why, then, is disability seen as a problem? As philosopher Elizabeth Barnes writes, disability is often characterized as a bad-difference, but she reminds us that it can be seen instead as mere-difference. What is bad is living in a world that tells you that you are wrong for existing. Many of us want futures where disabled people don't have to be pushed to normalize our mere-difference in order to make ourselves *palatable* enough—"includeable," to use sociologist Tanya Titchkosky's term. But in the current climate of technological solutionism, where my university hosts talks on "The End of Disability" and Herr brings his brand (he does want to sell you his ankles) of technological relief to MIT, disability is painted only as tragedy and is used as a backdrop for stories of tech-driven intervention and subsequent redemption.

Narratives about disability technology, in all sorts of media, do not serve the people the stories are ostensibly about. In fact, *the stories aren't really about disabled people at all.* These stories, filtered through abled imaginations, simply reinforce tired tropes about technological progress and techno-optimism and the progressive power of engineering. The people with disabilities are secondary characters; the real story is about the triumphant march of technological prowess, of better living through technology. And in turn, the more of these stories people read, the more they internalize these feel-good narratives that populate our media and sports coverage. The standard narratives about individuals "overcoming" disabilities tokenize certain types of disability while erasing others.

These attitudes creep into even the best-researched and best-written articles about technology and disability. A BBC2 story headlined "First Prosthetic Limb Designed for Women: 'I feel liberated'" is quite good; the BBC anchors introduce us to the "new, hi-tech bionic hand" created by the Bebionic company and one of its users, Nicky Ashwell. Unlike many examples of disability tech coverage, the article interviews an actual user (and Ashwell is brilliant here) rather than limiting its coverage to nondisabled engineers, parents, medical professionals, and so on. Nonetheless, the news segment still gets it wrong in many ways.

In an awkward scene, the reporter shakes Ashwell's bionic hand for a bit too long, not letting go, and asks, "What is it like, psychologically, to be able to shake someone's hand?" You can see the narrative the reporter is angling for in the questions asked: they want to hear that Ashwell's arm has changed her life and changed her for the better psychologically, that something that was "missing" has been restored or replaced. But Ashwell eludes that line, giving answers about function, explaining how useful the hand is as a tool. She even explains why other hands were not for her and how for a long time she didn't even use a prosthetic hand. That headline quote about being liberated is not about the Bebionic hand itself but about *riding a bicycle with the hand*. And who doesn't feel liberated on a bike? While the hand allowed Ashwell to easily ride a regular (not adaptive) bicycle, the feeling of liberation came from riding the bike.

Yet that title, "First Prosthetic Limb Designed for Women: 'I feel liberated,'" highlights the story they wanted to tell, despite Ashwell's own down-to-earth and practical descriptions of the new hand. She likes it, and it's great that there's a prosthetic hand option designed for women and appropriately sized, but she's not a different person with it, and it didn't liberate her from some bondage of disability. The hand is a tool that eases some awkward social situations, that makes tying lace-up shoes easier, and that lets her hop on any bike she likes.

Tropes of disabled people "liberated" by technology *stick*. The Bad Cripple, Bill Peace, writes: "These images of prosthetized amputees and walking wheelchair users—taken as a representative of disabled people—make for a highly problematic case of a tokenism that insists that disability is something to overcome, something to eliminate, something that is simply a medical problem and not a social one." Organizational psychologist Bertolt Meyer and social psychologist Frank Asbrock have written about how bionics—prosthetic arms and legs, exoskeletons, and retinal implants—change perceptions about disabled people. They study stereotypes and how they shift when it comes to highly technologized disabled people (cyborgs). In two online studies published in 2018, these researchers asked questions about how people using bionics were perceived in terms of competence and warmth. They found that cyborgs are perceived as more competent than other physically disabled people (though still not as competent as nondisabled people) but less warm.

I'm not averse to technologies. Obviously not; I use a prosthesis every day. When my fake leg is broken or ill-fitting, I care very much about getting it fixed so that I can engage in my normal activities in the way to which I have become accustomed. But other, less "advanced" technologies are equally important, and being familiar with other devices is essential. Some technologies can be more useful at certain times than others. My forearm crutches came in very handy when I took a fall a few months ago and was off my leg for five weeks, and I love my rollator for bopping around the house; it's the best on smooth hard floors, and I can go so fast.

What makes me angry is the huge gap between my experience and the stories in the media. There's such great divergence between the experience users have with advanced tech like prostheses and mobility aids, and the whiz-bang media coverage of these items. There's likewise a huge disconnect between how shiny new aids are promoted and how they are adopted.

Disability (a social category) isn't a problem to be solved by technology. But MIT Media Lab's Hugh Herr regularly says, "I don't see disability, I see bad technology." There's a tension here—between a disability community that is centered on disability rights, justice, and community pride and Herr's vision of disability as a technical challenge. Both can exist at the same time. But media attention and storytelling focus almost exclusively on disability as a technical challenge, and this undermines disabled people's lived experiences and community. Framing disability as *the* problem turns attention away from the real problem: the world is set up to exclude disabled people.

My friend Cynthia Funk, who I met at my prosthetist's office (she remains a great amputee mentor to me), put it just right in a Facebook post some years ago:

> Newer amputees see these events unfold in the media, and expect to be able to strap that leg on and go . . . We see amputees dancing away on television. Running marathons, simply walking. These fortunate few are the focus of news media. The bed of roses, inspirational stories that the general public is exposed to is far from reality.
>
> Please! Before you question or comment to an amputee why that person you casually run into at the grocery store or other public place isn't wearing their leg (or have one at all), or why they aren't doing certain activities, realize that they may be living a normal, realistic existence in a media-ridden society.

A similar statement is found in the writing of Donna R. Walton, founder of the Divas with Disabilities Project and advocate for disabled Black representation in entertainment:

I worked so hard to build and maintain self-esteem and confidence that I almost risked losing it when I decided that I had to walk "normal" again in order to attain society's ideal for physical attractiveness. My gimpy gait is mine, and it is very "normal" to me. I feel now that I have evolved into my own reality consistent with being a woman, an African American, and subsequently an amputee, with each level of my survival affording me enough clay to mold a self-assured persona unique to my struggles. . . . Just being able to walk or to get where we want to go, whether it is with a cane or in a wheelchair, is the point, isn't it?

The media shows us mostly successful, inspirational, white or lighter-skinned, good-looking, socially acceptable amputees—often war heroes or elite athletes. But these types of "good" amputees don't represent the majority of amputees. It's hard to speak with authority about exactly who makes up the group—for one thing, the data doesn't account well for congenital amputees who don't undergo surgical amputation. For another, the estimates of people who get amputations each year lag by a decade or more. The estimated numbers of "nearly 2 million people living with limb loss in the United States" offered by Amputee Coalition of America (which hosts Limb Loss Awareness events, amputee kid summer camps, peer visitor trainings, and an annual conference) are based on data gathered before 2010. According to these non-ideal estimates, of an estimated 185,000 amputations each year, causes range from "vascular disease (54%)—including diabetes and peripheral arterial disease—[to] trauma (45%) and cancer (less than 2%)." I'm in the less than 2 percent bracket of cancers with my osteosarcoma bone cancer.

According to the Amputee Coalition, African Americans have limbs amputated at four times the rate of white Americans.[12] *Pro-Publica* investigated this disparity and in 2020 ran a story by Lizzie Presser called "The Black American Amputation Epidemic." Presser writes: "It is the cardinal sin of the American health system in a single surgery: save on preventive care, pay big on the backend, and let the chronically sick and underprivileged feel the extreme consequences." The article follows the work of the only cardiologist in Bolivar County, Mississippi, Dr. Foluso Fakorede. This doctor works to save limbs and feet for patients whom other doctors wouldn't spend time on; he aims to save lives and give patients dignity in the face of an indifferent-to-malicious medical system that has long neglected them. This kind of story about amputees—and disabled people more broadly—is vanishingly rare. Perhaps the most unusual thing about Presser's story is its explicit attention to those disabled people who are most often forgotten or erased. Presser maps geographic areas where high annual amputation rates cluster and overlays it with maps that reflect the concentration of enslaved people prior to 1860, showing that amputation rates reflect long-standing disparities in healthcare. Amputation statistics tell us something harsh about who matters and who doesn't with respect to healthcare and survival. This part of disability reality is almost never mentioned in stories about disability technologies.

And this was before Covid-19, which has radically shifted what the amputee community looks like. Some of us were already in categories of "high risk" before Covid, with complications from other circumstances; many others have joined the community because of Covid-19. The disease itself has caused many amputations due to vascular complications, and hospital overcrowding during surges has led people to delay medical care (including wound care), potentially missing out on measures that might have helped doctors salvage their limbs.

Vascular complications, often from diabetes,[13] loom large in the world of amputees—and those who might become amputees. Half of the amputees from vascular disease, the Amputee Coalition tells us, "will die within five years," which is "higher than the five-year mortality rates for breast cancer, colon cancer, and prostate cancer." This last statistic really spooks people who enter our community, since they often learn about the higher mortality rates of amputees without any medical context. Amputations themselves don't necessarily cause higher mortality; people who have limb amputations due to diabetes and other vascular diseases have advanced forms of disease already, and this at least partially, if not largely, accounts for the higher death rate.

The common, omnipresent association of amputation with war heroism in the media actually has effects in our daily lives: we are very often approached in public and thanked for our service, especially male amputees with short hair and muscular builds. I was thanked for my service more often when my hair was very short (after chemo). I was closer to looking the part. My amputee groups have whole conversations about how to deal with these encounters. It's especially complicated for people who did serve in the military but did not receive their amputations that way. I have one amputee friend who was a mechanic in the Air Force but messed up his legs in a car accident. He dreads these "thank you for your service" encounters. He has real respect for people who were injured in the line of service, but he was never in combat situations. It seems easier just to say "you're welcome"—you shouldn't have to launch into a long personal history about your service and health to a stranger in order to pick up cabbage at the grocery store—but nobody wants to "steal valor" from those injured in the line of duty.

It's not that public images of amputees are never true—including the technologized images from *Dancing with the Stars* to TED talks on prosthetics. Many of us do seek to address some aspects of disability in technological ways, and techno-optimism is seductive. I adore my prosthetic leg (when it fits right); we go everywhere together. But I do not like being lumped into the "inspiration" category because I can walk with my prosthesis. This is an effect of the overriding media narrative about disability and amputees. I've been told I'm an inspiration while picking out corn in the grocery store.[14] While I don't enjoy getting into fights in the produce section, I also don't like being treated as a testament to the power of technology/the human spirit, or as a cautionary tale about risk and responsibility (the narrative often applied to diabetic amputees or accidental-injury amputees). So much of what I experience as someone using disability-related technologies is tied to my technologically embodied configurations, and the wear and tear of technology—in ways I've never seen addressed in popular media.

What do technologies mean to people? What role do they play in people's lives? How does the built environment include or fail to include people? How do people identify with and resist technological intervention? How (and when) do people decide to adopt or use particular technologies such as prostheses? These questions should not be the exclusive province of abstract philosophical speculation or techno-futurist disruptors. They must be grounded in the real experiences of real disabled people, whose lives and bodies are a special testing ground for these ideas.

When the media image of amputees on running legs becomes the only representation of what it means to do well as an amputee, it presents significant problems—not just for other amputees but for other disabled folks. This singular narrative (a form of tokenism) insists that being whole and healthy means being a heroic inspiration to others (even though sometimes it frankly sucks to deal with a disability, especially when you are newly disabled) and always projecting a positive attitude (otherwise you aren't really trying). It means asking us to live and perform as nondisabled people do, to do so without complaint or discussion of our complex and nuanced lived realities, and to serve as examples of what a "good" disabled person looks like. This media narrative doesn't let people be real or honest or express complicated feelings of grief or joy outside of the script our society has set up.

This prevailing narrative doesn't let us be *disabled* at all. Derek Hough, the professional dancer paired with Amy Purdy on *Dancing with the Stars*, showed the consequences of this attitude in a blog post he wrote before he even met Purdy: "For somebody to take the chance and be courageous enough to do a show like this, it's just really inspiring to me. . . . I could tell without even meeting her that she didn't see herself as disabled."

Even without meeting her, he thinks she "didn't see herself as disabled." Apparently, it's bad for disabled people to consider themselves disabled—even though our disabilities often inform every aspect of our lives, our experiences of the world, and our identities. This is the script. We couldn't possibly be disabled and want to dance. And I'm guilty here too: as I boogied with my kids in the kitchen before my amputation, I wondered if I would ever dance like that again; I had a very narrow concept of dance and of disability then too. It's different on the other side—and *being* disabled is easier than *becoming* disabled! But the idea we're given, especially by this kind of news and media coverage, is that if we dance, we are not properly disabled; we've removed ourselves from that category by *overcoming* through technology.

For new amputees the message is that you need the right attitude, enough spunk and pluck, to overcome (neglecting other health concerns and/or costs associated with therapy and technology). For other disabled people: if you aren't like one of these lucky amputees with high-tech legs or arms, your time will come, too, with enough research dollars and hope.

But dancing is for everyone, including disabled people, whether they are technologized or not. The film *Invitation to Dance* is about disability activist Simi Linton, who is also a wheelchair dancer. Linton describes dance as "the expression of joy and freedom." In this narrative, wheelchair dance and disability activism are shown to be extensions of valuing and appreciating disabled embodiment, disabled movement, and disabled people.

I am with Simi Linton. Dance is joy and freedom expressed through the body—*every* kind of body. The big event of every Society for Disability Studies conference is the dance. At my first SDS in 2015, I didn't know what to expect. What I found was more free and joyful than I'd dreamed it would be: My then-new friend Mallory lay down on the ground and danced with her arms and leg, and encouraged others to as well. A tall, blind white man with a white cane rode that cane like it was a horse—a signature move. There was quiet space outside the room for those who needed it, and a smiling autistic friend wrote a note to let us know she had gone non-speaking as a result of happy overwhelm, and so she couldn't respond verbally. At one point, a Black woman glided very regally in her wheelchair onto the floor as the music changed from upbeat to more lyrical, accompanied in dance by a short slim blond woman. Everyone moved to the side to watch them dance: this wheelchair user turned out to be dancer-choreographer Alice Sheppard (though I didn't know that at the time), and this was years before her Kinetic Light project's "Descent" performance and production. "Descent" consists of stunts and specialized ramps and lighting for wheelchair dance, but there on the dance floor at SDS I was no less awed. There was a moment when our DJ was about to take pictures of the crowd dancing, and about five disabled elders quickly surrounded him and told him no. Not only might flash photography set off seizures or migraines for some of our people, none of us want to be memed into inspiration, to have our images taken and used without permission, *to be the subject of others' imaginations rather than our own creation.*[15]

On *Dancing with the Stars*, amputee contestants are framed as courageous for showing up at all. Certainly, it is hard to be stared at (by millions of people, no less) while trying something you might do poorly. I don't doubt that takes courage. Yet the amputees we see on shows like this are well adjusted to their prostheses and have more access, wealth, and health than most of us. The typical media stories about disabled dancing gloss over the difficulty, discomfort, constant monitoring, small adjustments, insurance worries, and so on of the lived experience of wearing high-tech prostheses. In these stories, we are given an image and story of rejecting disability by overcoming it, rather than talking about how we celebrate disability in our movement, how dance is for every body. I want to celebrate the diversity of our bodies and movements and styles, as we did on that dance floor at SDS 2015.

While my prosthetist has many videos of amputees in his clinic showing off their new struts and climbing stairs, he's very clear in conversation about what to expect. For many of us, after we've been tweaked repeatedly and gotten a good fit on a socket, we are thrilled with our new legs, riding the high. But *it can't stay like this.* We do what I think of as "falling out of true." You know how we talk about "true" wheels—wheels that are perfectly balanced and aligned, perfectly fit for purpose? All mechanical objects (whether wheels or legs) also go out of true.[16] Our bodies change, and so does the fit of our devices.

This process of being in and falling out of true happens with all disability tech. Getting fit for a leg (or arm, or custom orthotic device) is an iterative and interactive process. We might imagine that a person goes to one doctor's appointment and walks out with a leg, and what happens after that is up to them. If they have the right attitude, they will make it work—and compete on *Dancing with the Stars*, or take up running. Weirdly, I was asked by friends whether I was going to take up running once I had my first prosthetic leg. I hated running before my amputation. Why would that change? And now I'd need to spend big money on a leg my insurance won't cover to do the thing I don't like? No, thanks.

How it actually goes (though, of course, it varies for different levels of leg amputation, whether at or above or below the knee or hip): after the amputation, before getting a prosthesis, you first have to heal up from whatever type of amputation you had. During this time, you generally begin reaching out to prosthetic offices to find someone you'd like to work with. Sometimes prosthetists meet with new amputees in hospitals, and other times there are recs or referrals from surgeons; in many cases, there are limited choices near where you live. During this period, people are sometimes doing physical therapy to help with core strength and hip flexion to prepare for the prosthesis. Many new amputees are disappointed to find that you usually don't walk out of your first appointment with a new device. The processes can vary by method, but you will be spending some time there and coming back. Different systems and components have different issues and advantages. Some people's limb volume hasn't stabilized; for these people, vacuum systems don't maintain vacuum very well. Even people whose limb volume has stabilized experience limb changes during the day. Just like your feet may swell in your shoes with salt intake, hot weather, menstrual cycles, and more, your whole limb changes circumference throughout the day.

For people with higher amputations, there are prosthetic hips and knees—components that prosthetists buy to be part of your leg's assembly. Like feet, they are sold by different companies, and like feet, people hopefully get to try out different hips or knees to see what works best for them. Above-the-knee amputees need their computerized or hydraulic knees to be configured and programmed, often multiple times during the test-fitting process. From our amputee statistics, we know that 70 percent of leg amputees are below the knee, so feet are a big market. The next most common type of amputations are above-knee and knee disarticulations. With the right insurance, one's prosthetist can choose, hopefully with your input and ideally trying out some options, from a variety of feet and knees. (Although many amputees are not given options by their prosthetists, they should be. And it does take a prosthetist ordering a bunch of components they know they will have to return, so it can be a hassle. Feet are in specific sizes and sidedness—left-right—so most prosthetists don't just have a fun closet full of feet and knees to let people try on.) The smallest group of amputees are hip disartics and people with hemipelvectomies, and so there are fewer options for hip components. HD/HPs, as they are sometimes abbreviated, need to select, fit, align, and test all three joints—foot, knee, and hip.

Most advancements in prosthetic legs trumpeted by the media are not about sockets but about feet, ankles, and knees. But none of these other prosthetic elements matter if the socket doesn't fit well enough to be comfortable, which is a matter of human skill and human variety, not just technology. The process of iteration for fit is absolutely crucial, because a badly fit socket or poorly aligned limb can cause tissue damage, blisters and other skin issues, damage to muscles (and not just on your leg: we are talking about something that uses and changes your posture and gait), all of which incur costs in time and safety (and expense and lost wages and extra appointments). People often wear painful legs at risk to themselves—or don't wear uncomfortable legs, simply sticking them in a closet somewhere. A badly fitted socket increases the risk of infection through skin issues and even reamputation—what is called "revision," where a surgeon would cut you again above your initial amputation site in hopes of getting a better situation for your prosthetic socket. And, of course, the final or definitive socket isn't really final—you will need other sockets, because you'll wear out every bit of your leg if it's a good one. Even with a great socket, you'll bust feet, wear out components (like the squeak-dampening spectra socks that go between foot shell and foot), and more. And at some point, your great final socket won't be great. It won't fit right anymore. It'll be time for a new one. Lengths of time between legs can vary widely, with newer amputees needing legs more often as their bodies adjust. There is no standard timeline for getting a new socket or leg. One of my colleagues had the same leg for twenty years and just got a new one; he's a significant outlier. Another friend was in three newly cast test sockets in the past year alone as their limb drastically reduced in volume; thank goodness she wasn't rushed to take a final socket too soon.

All these things are expensive. A C-leg (a type of computerized knee) costs around $50,000. This is before it's attached to the socket or foot. Normal, unfancy below-the-knee prostheses, without any computerized foot/ankle components, run between $8,000 and $16,000. Hip disarticulations and other high, above-the-knee amps require bucket sockets and belt systems to keep a leg on; my friend Mallory has been quoted $100,000 for a new leg. We all just have to hope we have good insurance coverage. Medicaid and many programs require amputees to pay 20 percent, meaning that a person's leg, even with insurance, could easily cost more for them than a used car would, especially for higher-level amputees.

In the tech community, feet get a lot of attention—but many of the hot new feet with computerized ankle components (like the BiOM ankles that Hugh Herr wears) are heavy, and some prosthetists believe that they are not worth the physical cost of wear for people without adequate core strength. Often tested in VA clinics, they work best on athletic males who were sporty before their accident or injury, like Herr himself, a world-class mountain climber. The women I know who have tried them have all moved back into more static carbon-fiber ankles. Having a more expensive part to your prosthesis doesn't mean a better gait if you don't have the musculature for the added weight. This tech is a "solution" for far fewer people than hyped.

There is generally a different relationship with tech among arm amputees than leg amputees, and there are a lot more leg amputees than arm amputees. Many prosthetists have less experience or are ill trained on arm prostheses. Many more arm amputees than leg amputees choose not to try prostheses at all, or decide not to stick with them for the long term. Depending on the level of amputation, carrying something extra around on your arm may not be comfortable, and arm components have often been clunky and awkward. Indeed, many prosthetists see mostly leg amputees and have good experience in making legs. Arm amputees need to be extra careful to interview any prosthetic outfit to see what experience they have with arms, since this is not a given for every prosthetic clinic.

The fact is that arms and hands do a lot more things than legs and feet. Legs and feet are mostly for walking and sitting down: those are the main tasks for prosthetic legs. But arms and hands have more complicated jobs. Right now, new technologies in arm prostheses, which offer a wider set of options, are making the news. One area that we see a lot lately is 3-D-printed hands on children, with lots of projects about "hero arms" and other colorful and affordable devices made outside of normal prosthetic clinics—and sometimes in libraries and other very non-medical contexts with 3-D printers. Ideas about what arm prosthetics can be and look like are also starting to open up. In my previous work in animal tool use, I've seen animal-studies scholars compare human (and raccoon) hands to "multi-function tools," but some of the best hand prostheses are made for specific functions: for example, I love the tattoo-gun hand that made the news a few years ago. It allowed the wearer to get back to the work they loved with a hand that they themselves helped design to meet their needs. In her book *Staring*, Rosemarie Garland-Thomson notes that functional hands, like hook tools (or tattoo-gun hands), "answer the needs of the wearer," while some arm and hand amputees wear cosmetic hands "to answer the needs of the starer." Having a five-fingered hand is seen as "normal" by the public, just as walking with a good gait is seen as "normal functioning" for leg amputees.

The idea that five-fingered hands equal normal functioning is the foundation of most media coverage about arm and hand prosthetics. Achieving that vision of normal is seen as life-changing for the disabled person—despite the fact that five-fingered prosthetic hands aren't always that functional. As Jen Lee Reeves, mother of a daughter with limb difference, writes in a blog entry entitled "Prosthetics Do Not Change Everything,"

As a mom who is the parent of a prosthetic-wearing limb differ-ent child, I know the reality. . . . We've had moments when she did something awesome she couldn't do without her helper arm and it has brought me to tears. But those are moments. Moments Jordan wouldn't have enjoyed without the help of prosthetics. I guess you could say they changed Jordan's life. But prosthetics don't make everything perfect. . . . Prosthetics aren't a solution. . . . I believe in the use of prosthetics as a tool. 3D printed hands and arms are very cool. But they don't change everything in the life of a limb different child.

This entry was written in 2015, and Jordan is much older now. I'm in contact with Jen through disability-studies circles, and Jordan has gone on to open up space for thinking differently about arm prosthet-ics for children. For one thing, as part of Project Unicorn, she designed her own arm, based on a design she made when she was a kid. It looks like a unicorn horn and shoots glitter. It's not an everyday arm, but it is the arm she imagined as a child, outside of the "normal" paradigm of the five-fingered arms we see making news. These days Jordan often chooses not to wear an arm at all.[17]

Right now, amputees have choices that other disabled people don't when it comes to how we want to be in the world[18]—whether we want to use prostheses at all, what kinds of prostheses we want to use. But these choices are always made within the context of social pressures that push us toward normalcy, inviting comments and critique.

For leg amputees and many others who may use leg orthotics, walk-ers, scooters, and other mobility equipment to ambulate, the pressure to walk and to walk well is no joke, so deeply embedded are notions of fitness and vigor to being a good citizen and person. I think everyone has some internalized ableism surrounding walking—it's one of the difficult things for amputees to root out of their own heads and for people to address in their own impressions about the competency and respectability of others. Bill Peace, who taught in Syracuse University's Disability Studies Program, blogs about our cultural hang-up with walking as the gold standard for normality:

> Your typical bipedal person exposed to a barrage of misleading news stories is led to believe all paralyzed people share one goal in life—walking. Please cue the soaring inspirational music accompanied by the brave and noble young man or woman struggling to walk surrounded by health care professionals, computer scientists, and engineers who share the same ritualized ideal.

He continues, just to amplify earlier mentioned of models of disability here:

> The flip side of the obsession with walking is not discussed. No one wants to talk about the gritty reality people who cannot walk are forced to navigate. No one wants to think about the barriers to health care and appropriate adaptive technology. Paralyzed people who want to be ordinary are a problem. They are stigmatized for their failure to follow the societal script. Worse, they force others to obey the law, the ADA in particular, that prohibits discrimination against people with a disability. Like typical others we people with a disability want to be able to work, own a home, and have unfettered access to mass transportation. For this to happen disability must be placed in the larger societal, historical, political, and medical context.

You'll recognize in this entry the tension between the medical model, in which disability is the problem, and the social model, in which the built environment is the problem. It's very clear which model the tech world subscribes to, with its expensive, complex solutions to the "problem" of not being able to walk. I think often about how these technological solutions affect the way disabled people experience and think about our bodies.

Jai Virdi, in the historical account *Hearing Happiness*, and Sara Nović, in the novel *True Biz*, write in a similar frame about expectations around hearing and how they shape how individuals think about themselves (and push back). The recent history of cochlear implants—and indeed earlier hearing devices too—is an excellent case study in the ways "helping" and medical professionals have significantly shaped the dialogue and public understanding of deafness and pushed an optimism for technology that often is not cleanly experienced by the objects of the helping.

Cochlear implants are viewed by the general public as "curing" deafness, and parents of deaf infants are often strongly pressured by medical professionals to make early decisions about their child's hearing, including whether to surgically implant them with the devices. One of the first things a newborn in the United States receives in the hospital after birth is a hearing test, with only two possible results: pass or "not pass." And the test will be performed again, but a "not pass" raises early worries and will lead to further testing. Not every deaf child is a candidate for cochlear implants, and not every surgery is successful. And most new parents don't know that the Deaf community resisted cochlear implants, or why. Videos of audiologists and medical professionals turning on a child's implant for the first time—where a child jerks their head in the direction of the sound of their mother's voice—have gone viral, serving as the only introduction to the subject for a wider public. These viral videos praise, sometimes explicitly but certainly implicitly, the power of technology to end disability. People reply in the comments with "praise hands" emojis and gush about the power of technology and thank God.

But the videos don't give any clue that not all Deaf people consider themselves disabled, that the child in the video will need years of interventions to be able to "hear" and communicate well with the cochlear implant, that having a technology for disability (even when it works) still means a great deal in terms of maintenance (batteries being sort of an everyday kind of worry, along with others), and that implantation removes any natural hearing a child may have. This technology story—highlighting just a few seconds or minutes of a surprised child (usually they show the videos where the child smiles, not screams or cries)—is given to viewers without context of the wider implications or any sort of follow-up and without any information from adults with cochlear implants about their various experiences.

Nović's novel *True Biz* is about a Deaf boarding school, where students learn or know American Sign Language, and the book lets us look through the lens of several different characters with different positionality and perspectives in the community. Some of the teens in the book are from Deaf families, some are from hearing families, and some have used different hearing devices. A central character, Charlie, has been raised by a mom seriously committed to Charlie being "normal," to Charlie being hearing—discouraging her from looking like she might need help and from adopting other modes of communication. Charlie grapples with and tries to understand her mom's perspective, influenced by a medical establishment that told her that these were the right things to do for Charlie's success.

The Deaf community is not technology-averse as much as they get painted that way. In the novel, Deaf people use video phones to talk to parents in ASL and use computers and cell phones in all the ways we would expect from average teens. Too often, our narratives about hearing technologies—throughout history, if you read Virdi's *Hearing Happiness*, which looks at historical advertisements of hearing technologies—proclaim miraculous cures. The people who receive the technology and remain unsatisfied or who refuse to use it or hype it up are cast as not sufficiently grateful, as Luddites, and as stubborn fools.

A few years ago, someone complained to me that my grandpa didn't bring his hearing aids to the restaurant where we were about to eat, revealing how little they knew about the experience of technologically assisted hearing. I have a set of hearing aids to address being hard of hearing after chemotherapy (one of my drugs was oto-toxic; not all chemo will do this to you), and I learned early on to take my hearing aids off before dinner. In fact, I had already put my hearing aids away in their case at this point because the amplification of chewing noises cuts significantly into my enjoyment of a meal. And they make me hear actively worse in environments where many people are speaking at once and from different directions. Even in what others would call quiet, the amplification of so many whooshing noises of air conditioning and vacuum cleaners and outside construction is overwhelming.

It can be exhausting to have to hear so much. Back when I would use my hearing aids more regularly, I would look forward to the time of day each day where I could take them off (along with my prosthetic leg, decyborgifying for the day). Some people in Deaf Studies talk about Deaf *Gain*—in opposition to hearing loss—to refer to the benefits of being deaf and what they have gained from both Deaf community and from not being able to hear. Too often, we have been conditioned to associate deafness with loss and to think that people need immediate technological fixes. But even technologies that can be helpful are not the simple replacements they are imagined to be. Things don't work the same, or even as expected.

The push for automation in bionics—of wheelchairs and of fake limbs and, well, of bodies more generally—changes not only how people are thought about by others but how people think about themselves. The technologized disabled body—the re-enabled body, "triumphant" over its own conditions—is a lie. Technology cannot transcend the meat sack; the body is still there, still felt, still handled, enduring. But technology—and the normative ideas of what it means to have the *correct* body or mind—increasingly separates our *selves* from the bodies with which we encounter the world. Like some people with acute and dire medical problems who talk about feeling alienated from their physical/embodied selves, disabled people often need to (re-)integrate our selves with our bodies, while living in a world that instead tries to force us into a fake normality.

While support can now be found in online groups, there's a very important materiality to prosthetic use that is best understood through in-person contact—seeing other people like you, making those personal connections with people who understand. The clinic or physical therapy office or anywhere else we can congregate, like an Amputee Coalition meeting, should be a place of empowerment, giving you tools to navigate the world in familiar ways again, but the tools aren't only technological or about bootstrapping up. Only other disabled people can understand what it's like to live with this type of tech so closely: that our devices are often annoying and squeaky, that sometimes things hurt, that sweat is hard to deal with. The bodied experience of prosthetic and orthotic fitting (and wheelchair fitting, for that matter) and community therein is already obscured in our media, but it's also being erased as a process—fading from existence due to increased automation and conglomerations of prosthetic facilities and equipment dealers. (The wheelchair repair crisis is already upon us, with pending legislation around the "right to repair" that could make a difference.) As old tech is superseded, mom-and-pop–type prosthetic facilities are bought out by larger corporations, which very often serve patients less well and certainly intensify the drive for more and more expensive tech. There is profit to be made, after all.

New technologies raise questions about the sorts of people we want to be and the sort of society we want to live in. This is true for many types of technology, not just disability-related technologies, and some communities are more intentional about critical consideration, periods of assessment, and decisions about adoption. Rarely do we see disability technologies involve the kind of reflection raised by other technologies—even though these are often technologies we wear closer to our bodies, often in everyday use, ones that need maintenance and repair.

ZERO—

I'm an amputee, no longer becoming but being. My spouse has learned all the ins and outs of our insurance network's plans; that's the painful part. I head back in a few weeks to visit my surgical team after not visiting for over two years (Covid). They'll write me a prescription for some new prosthetic liners. I will just see the nurse practitioner, who has written a lot of these prescriptions for me over the years, since the surgeon has more pressing cases to see. My prosthetist will use his repurposed pizza oven to shape new liners over a cast of my leg, then ship them to my home (his office is great to do this!), and then I'll start breaking them in. I don't think we'll be talking about a new leg until next year, because this one, two years old, is still feeling right—even though it feels better than snapping off a bra each night when I transition to no-leg o'clock, when I go really fast on my rollator as I glide from bedroom to kitchen to fix an evening snack.

CHAPTER 5

The Neurodivergent Resistance

■　　■　　■　　■　　■　　■　　■

T HIS CHAPTER IS "OUTSIDE MY LANE." I'm not autistic, but I'm writing here about techniques and technologies for autism specifically, and about neurodivergence more broadly. As a disabled person in cross-disability community with others, I pay close attention to conversations and advocacy from the autistic community. What they are doing matters to me. After I acquired most of my disabilities in my early thirties, I came back to a campus and workplace that seemed incredibly hostile and sought out a community. Other disabled people, of all ages and experiences and with all kinds of disabilities, were there for me.

What do I mean by incredibly hostile? I was coming off a year of chemo, much of it inpatient in hospital. I had no stamina, no energy; I was on forearm crutches with what felt like a very heavy prosthetic leg; I had a million doctors' appointments all the time. I was also adjusting to so many new disabilities—an amputee, yes, but also now hard of hearing with constant tinnitus and chemobrain, and with hot flashes, too, after having been sent through a very rapid menopause on chemo.

Oncologists kept saying that my body had some recovering to do and that we'd check after a year to see if my tinnitus went away and if my hearing improved and if my menstrual period would come back and if my cognitive impairments would improve. I would not get any of these things back.

But my chapter here is not about my bodymind alone. It's about what was all around me. Every space that once I breezed through without noticing was transformed. Getting to the building I worked in felt impossible some days: on my campus, even close ADA parking was not close. My office had been moved downstairs (away from my colleagues) since our building has no elevators. My tinnitus and hearing changes meant that places sounded different; noises intruded on my concentration (what there was of it) in ways they hadn't before. I was also stared at wherever I went: a bald amputee with crutches tends to draw the abled eye, and I moved slowly enough that I couldn't get away from the stares. Even though I had the absolute best colleagues—they brought meals for my family when I was sick, offered childcare, and were as supportive and kind as people could be—just being on campus was very hard. My bodymind was no longer the type of person who could feel welcome and at home on campus: *every place was unprepared for me.* The entrances to buildings I needed to use were all on the back or the side, impossible to find. There were stairs everywhere. And thanks to chemobrain, I now needed a lot more in the way of reminders and communication than I did before. Even the best and most supportive nondisabled colleagues couldn't really understand how alienating it was just to be there. Thank goodness I knew some disabled people. They got it.

There were six key early neurodivergent members at meetings of what became the Disability Alliance and Caucus at my university.

Three of our six were autistic (and had other disabilities too). One didn't have a diagnosis yet but was definitely experiencing symptoms that needed accommodations that were not being provided. One was a cane user due to a neurological condition. And me, in all my multitudinous glory—amputee, tinnitus, hard of hearing, chemo brain, Crohn's (that would be diagnosed soon after), possibly some acquired medical-related anxiety just to give me that extra sparkle. We were at all different life and career stages—two undergrads, three grad students, and an early career half-time faculty member (me).

This kind of disability community is at the core of the current wave of disability activism and theorizing. A lot of the hottest current work centers the idea of *neurodivergence*, and it's coming out of a budding and overlapping coalition of disability communities, much of it from communities of autistic people and people with ADHD. So many more adults have autism and ADHD diagnoses that they wouldn't have gotten a few decades ago. Back then, autism was underdiagnosed across the board; we didn't know then what it looked like in adults or in women or understand the many ways it can manifest in children. This is still true, to a large degree. Contemporary diagnosis for autism relies on older research that centers white boyhood. Although the idea that autism is primarily a disability of "ultra-male" brain has now been thoroughly debunked, it still has a huge influence on how autism is diagnosed and on perceptions of what it is. So we still miss autism in a lot of children, especially girls and kids of color. Things are even worse for adults. Diagnosticians still want to interview or survey people's parents and teachers—interviews that may not be appropriate or available for autistic adults. The current literature calls it autism spectrum disorder (ASD). This framing provides a way of making sense of the *many* spectrums of function and experience among autistic people; it is not one spectrum of function! Autistic people (and people with ADHD, who are also part of this history) have given a great gift to disability organizing, self-understanding, community, and activism with the ideas of *neurodivergence* and *neurodiversity.*

The "spectrum disorder" includes, for example, what used to be called Asperger's syndrome—a label that is no longer used diagnostically, although some people who got that diagnosis early in their lives do continue to use it to refer to themselves. Some autistic people are pushing to discard the name altogether, because Hans Asperger, after whom it was named, was a Nazi collaborator. He sentenced most people he identified as autistic to death or institutionalization, selecting only certain specific types of autistic people—those who presented with what would later be called Asperger's—as worth saving, as "real" citizens, because they could contribute to society (recall that the historical origin of disability was "anything that kept people from participating in labor"). I'll swing back to this idea later, but let me just give you a content warning: when we talk about the role of technology in autistic lives, we will see some very dark stuff. The historical treatment of autistic people (or people we would now understand as autistic) is filled with brutality and misunderstanding. When we're talking about tech for autism, we have to talk about both fidget spinners and gas chambers, about Dungeons & Dragons and locked closets.

In the early meetings of the Disability Alliance and Caucus, we were learning to work with one another and learning cross-disability information important to our self-understanding and advocacy. Undergrad founder Phoenix set us up with leadership files, name badging (based on ASAN guidelines), and information from knowledge they had from autistic-led college camps and training. We still rely on this organization—a balm for my chemobrain. It also was instructive for us on how to learn from one another in organizing and to look back on our records and agendas. Our group is still going strong, though our initial members have all graduated, forming and re-forming as new students, faculty, and staff lead us and guide us in understanding. Most of us were not raised in community with disabled people. Even those who grow up as disabled children are often not raised in the community: ableism means here that disabled kids are often encouraged not to associate with other disabled people. Most nondisabled parents and caretakers want to raise children to be as "normal" as possible. Indeed, many of us don't even know the disability community exists until we are disabled. Most of us find it because we go looking for *our people*— the people who get it, who understand our experiences and will share their experience and knowledge.

Phoenix introduced me to the terms *neurodivergence* and *neurodiversity*. Neurodiversity (sometimes called the neurodiversity paradigm) is, at bottom, simply a description of fact, an observation: brains work differently, whether the difference is in how we process information, read, react to sensory stimuli, or think. The term *neurodiversity* was coined by autistic researcher Judy Singer in 1998 and was originally conceived in relation to autistic brains. As professor Nick Walker has described it, the idea that "there is one 'normal' or 'healthy' type of brain or mind, or one 'right' style of neurocognitive functioning, is a *culturally constructed fiction*, no more valid (and no more conducive to a healthy society or to the overall well-being of humanity) than the idea that there is one 'normal' or 'right' ethnicity, gender, or culture."[19] Its sibling term, *neurodivergent*, was coined by Kassianne Asasumasu in the 1990s to describe a wide variety of diagnoses/brains/neurotypes, not only autistic ones; it is a more capacious category, intended to take a wider view on difference. The label neurodivergent includes psychiatric diagnoses, learning disabilities, brain injuries, ADHD, and cognitive disabilities of all sorts—any brain that doesn't think in "conventional" or expected or "neurotypical" ways. In the early days of the Disability Alliance and Caucus, this language allowed us to think both about being neurodiverse (as in, our group contained many neurotypes) and about how each individual one of us was neurodivergent in different ways: neurodiversity is about the aggregate (an individual person cannot be diverse, only a group can be), and neurodivergence is about the individual with respect to an aggregate (people are divergent from some standard or norm).

In addition to the neurodiversity paradigm, there is also a neurodiversity *movement*. The neurodiversity paradigm is simply observational—a statement of the fact that some patterns of thinking are different from others. The neurodiversity movement is a material application of the neurodiversity paradigm that pushes for civil rights for neurodivergent people—the respect and full social and civic inclusion of all neurotypes. The movement says: *Not only is it a fact that we have variation in how people think and process information, but we should value this diversity of thinking/processing/experience and make space for the existence of us all.*

Ideas like neurodivergence and neurodiversity build on the social (rather than the medical) model of disability, and they allow people to talk about different types of neurodivergence precisely as disabilities. To say you are neurodivergent is to align yourself with the disability community. When we understand the social components expressed in the neurodiversity paradigm, we can see how the environment that makes disability salient (or something that is noticed as a problem or issue, or made into a barrier): social norms and expectations make neurodivergence disabling, just as different architectural configurations make physical disabilities salient when moving through a building or landscape.

Think about eye contact. Many neurotypical people insist that eye contact is a marker of trustworthiness, while lack of eye contact indicates that someone is lying or ashamed. This is a cultural view of eye contact; while North America and Europe value direct eye contact, viewing it as a sign of understanding, engagement, and truthfulness (and, conversely, the lack of eye contact as rude, unprofessional, and a sign of disinterest), other cultures view direct eye contact as rude or aggressive. The expectations get built into technological systems: if you can't do eye contact as expected, online proctoring systems or other systems that monitor eye movements to track things like attention, truthfulness, and so on may flag you for non-normative eye contact. There are a host of problems with things like ProctorU and other AI software meant to catch people cheating—not to mention the invasion such apps represent. Recently, there have been additional objections to ProctorU's poor security, because the company has a lot of footage of people taking tests, including some people who cry—another thing that has been flagged as cheating.[20] All of this assumes that watching people's eyes is a good metric or indication of anything—but even if we were talking about only allistic people (that is, non-autistic) and assuming accommodations for blind people, eye tracking is pretty unreliable.

Some neurodivergent people struggle with this North American cultural and social norm for eye contact; for many (including many autistics), eye contact is painful and distracting. Even when people do try to comply with the social norm, there are many ways to "do" eye contact wrong—too much eye contact (staring) and unblinking eye contact are often read as hostile or awkward or creepy. Not all autistic people have trouble with eye contact norms and expectations; eye contact is not a good or reliable indicator of who is autistic and who is not, but many do struggle with this completely arbitrary norm that so much is projected onto. I think here of a blind friend who deliberately learned to move her eyes toward whomever was speaking because it made people think she was engaged with what they were saying—but of course she was no more or less engaged. As we saw with prosthetic hands, compliance with social norms is usually intended for the comfort of the nondisabled. For some autistic people, eye contact not only doesn't indicate engagement, it can be actively distracting and/or taxing.

I remember one meeting with a student who was out to me as autistic. She asked me if we could "do less eye contact today." Maintaining the facade of eye contact (a form of what autistics call "masking"—meeting social norms in order to make neurotypical people comfortable) was draining, and she wanted to be able to focus on our work. So, we looked in opposite directions and had a productive conversation about some mutual research. The eye contact was never necessary.

The idea that there's one kind of brain that is "normal" and good is, as Nick Walker writes, "a culturally constructed fiction." When we think in terms of neurodiversity, we create a big tent under which people with all kinds of learning, cognitive, developmental, mental health, and intellectual disabilities can work with the similarity of our differences, recognizing the structures, expectations, and norms that we can change, reset, bend, or expand to make a more inviting world for people with varying ways of being in and processing the world.

While the concepts and theories of neurodiversity and neurodivergence have come primarily from autistic people and people with ADHD, these concepts have been taken up and shared in and around the disability community. For some, the language of neurodivergence is used to push back against being considered disabled: the social model of disability allows the "misfit" to identify themselves with the nondisabled world, locating difference only in the mismatch between their reality and social and communicative expectations. But for many other neurodivergent people, it's crucial to see themselves as part of the larger disability community.

There's a reason people in disability studies like Margaret Price, Ellen Samuels, and Sami Schalk talk about *bodyminds*. Just as "purely" physical, bodily disability shapes our experiences and minds, neurodivergence, and the trauma that can follow it, affects our bodies. Neurodivergent people may feel physical pain and exhaustion from masking or from trying to process information given in tricky formats (as when someone with an audio-processing disorder or tinnitus tries to work with spoken information). Many of us also carry multiple diagnoses and other chronic conditions, and some physical disabilities and conditions are correlated to neurodivergence—for instance, autistic people have higher occurrences of Ehlers-Danlos Syndrome (EDS, a collagen disorder), Postural Orthostatic Tachycardia Syndrome (POTS, a condition impacting blood flow and balance), and digestive issues with related gut pain.

Disability is incredibly variable, both individually and community-wise, and disabled people sometimes make unexpected cross-disability connections that enrich our lives. This connection is often not fostered in clinics of physicians or therapists, which focus on individual limitations. Even when things like support or peer mentor groups exist in relation to clinical settings, they are often targeted toward one disability, not toward cross-disability understanding, and led by nondisabled facilitators who may not explore topics important to the group. Our variability and community connection are underappreciated among disability service professionals. These groups rely on pathologized approaches to disability (meaning focused on impairment) that can be in contrast to experiential or relational approaches (which would focus on environment, expectations, relationships, and set norms). Some forms of neurodiversity are framed as disabling only because of social mechanisms (like expected eye contact or not fidgeting too much). Other facets of neurodivergence are disabling beyond a social model and remain present and disabling even if social structures are changed to include a diversity of neurodivergent people.

That being said, addressing social structures can significantly improve the lives of disabled, neurodivergent groups:

> where arbitrary barriers are removed;
>
> where expectations are clear, flexible, and varied to meet different needs;
>
> where all the fluorescent lights are dead (replaced by something that doesn't trigger migraines) and there are no flickers;
>
> where things are quiet but with the right amount of white noise, and noise dampening headphones are available and socially unremarkable to don;
>
> where fidgets are allowed and people can move, sit, or stand in ways that work for them;
>
> where deadlines are flexible;
>
> where elevators are installed and working, or made unnecessary by design;
>
> where we are ramped up and hand-railed everywhere;

where food is labeled clearly, and cross-contamination
   hasn't occurred;

where everyone is allowed on amusement park rides and can go
   to concerts, and can be part of other entertainment and fun
   without being triggered without warning;

where there are places to go for downtime and rest;

where there are options for captions and image description—
   for live events too;

where gender inclusive bathrooms[21] are abundant and have
   adult changing tables;

where we can go cameras-off on Zoom, and have
   online options;

and so on.

But even with these social model–inspired changes (many of which
are part of universal design)—many people will still struggle. Some
will be in pain (though maybe with accommodations they will have
less pain, or get the time off they need when they have flares). Some
will have focus or attention that shifts and need flexibility in deadlines
or work processes. Some will still experience exhaustion and fatigue as
part of their lives, need technologies, formats, or techniques for communication, or need different forms of care to do some tasks (like eating,
bathing, cleaning). In other words, people won't become nondisabled
through the social model—it just changes where we seek reform.[22]

Because ableism has often been a central feature of conversations
around bodily autonomy, reproduction, immigration, education, inclusion, and even public services, and more intensely around additionally

minoritized groups, it's critical to understand disability as both political and relational too. Attitudes about disability often underpin how we consider important social issues that aren't even explicitly about disability: conversations about representation, abortion, welfare, voting, crime, and more are shaped by biases that often ride on lines about disability—who is best to make decisions, whose lives are valued, what supports people deserve in education and in the community, what/who emergency management plans for (and who will be left to die in an emergency), and how policing, social work, and public health get defined and enacted. We can learn from the past in terms of how disability has been politically framed, understood, and weaponized and see how our current systems perpetuate long-standing ableist biases. While social model work and universal design can make places more livable and accessible, real long-term inclusion of neurodivergent (and other disabled) people requires us to address the laws, beliefs, and policies that have created and reified injustice.

Pathologized or medical approaches to neurodivergence have often wrought violence on communities. There is a long history of eugenics being used against neurodivergent people, including those with autism and other types of highly stigmatized neurodivergence—especially specific types of mental illness, like schizophrenia, bipolar disorder, and various identity disorders, as well as developmental and intellectual disabilities of all sorts. Across much of North America during the 1900s, the most common way of "dealing" with many types of neurodivergent people was institutionalization and forced sterilization, with poor, Black, and Native people particularly targeted.[23]

Narratives about "defectives," "the feeble-minded," and undesirables fed many deeply ableist and violent social movements in the first half of the twentieth century, with legacies that are with us today.

We've borne witness to myriad dehumanizing narratives about disabled, particularly neurodivergent, people as "useless eaters" (a Nazi term) and "degenerates" (in the United States). Both the USA and Germany explicitly sought to improve the genetic pool of humanity and produce the best offspring. Eliminating disability, particularly congenital disability, was part of the "war against the weak" (the title of historian Edwin Black's account of America's eugenics campaign). Rooting out degenerate genetic materials and keeping them from being passed on was an accepted social and political project—one that was considered a "progressive" cause. Eugenics campaigns did not only target disabled people but all sorts of "others." They sought to eradicate anyone seen by the dominant group as genetically weak or inferior (including Jewish people, Native people, Roma people, Black people, and queer people). The idea that certain populations were more prone to disease, had low IQs, or were less productive provided a framework and logic for targeting people, a rationale that would justify promoting certain social and technological programs to intervene.

Eugenics was enabled by the latest technology. As Edwin Black has written, IBM (yes, that IBM, the one we know today, the one that currently has a training program for autistic workers) partnered with Hitler during the Holocaust to help exterminate undesirables. Indeed, "Nazi Germany [was] IBM's most important customer outside the U.S.," Black writes:

> IBM and the Nazis jointly designed, and IBM exclusively produced, technological solutions that enabled Hitler to accelerate and in many ways automate key aspects of his persecution of Jews, homosexuals, Jehovah's Witnesses, and others the Nazis considered enemies. Custom-designed, IBM-produced punch cards, sorted by IBM

machines leased to the Nazis, helped organize and manage the initial identification and social expulsion of Jews and others, the confiscation of their property, their ghettoization, their deportation, and, ultimately, even their extermination.

Among the items counted and cited by IBM data was information about ages [at which people had certain] viruses, racial characteristics, disease, and disabilities (such as lupus, syphilis, diabetes, influenza); IBM engineers and Nazi welfare experts used Hollerith Punch Cards to tabulate and sort this information. In *War Against the Weak*, Black quotes IBM's German subsidiary leader, Willi Heidinger, speaking to a group of Nazi officials in 1934 "about what IBM technology would do for Germany's biological destiny":

> The physician examines the human body and determines whether. . . all organs are working to the benefit of the entire organism. . . We [IBM] are very much like the physician, in that we dissect, cell by cell, the German cultural body. [This is a reference to the punch card system for sorting.] These characteristics are grouped like the organs of our cultural body, and they will be calculated and determined with help of our tabulating machine. . . . our Physician [Adolf Hitler] can then determine whether the calculated values are in harmony with the health of our people. It also means that if such is not the case, our Physician can take corrective procedures to correct the sick circumstances.

People who were seen as unproductive or deviant, as useless to the good of society and/or actively bad for society (concepts that depend on the notion of disability) were eliminated or subject to all sorts of other harms, including medical experimentation.

Other eugenics programs took a "positive" approach, employing a carrot rather than a stick. There were Better Baby and Fitter Family contests at state fairs in the United States in which doctors and nurses judged babies or families on their "fitness" on measured scorecards. Families had to submit their family trees to show how genetically good (and therefore morally "upstanding"—note the connection here between morality and the language of bodily posture) a family was. These positive eugenics programs encouraged "good" families to produce more babies. The flip side of this, of course, was the panic about those of lesser genetic and moral standing reproducing, causing the spread of involuntary sterilization. The Third Reich in Germany, with its concentration camps and gas chambers, is the ultimate expression of this logic. This is the beginning of the diagnosis of Asperger's syndrome: they were seen as good autistics who could contribute as citizens, who had higher IQs, who might be of value. Unlike other autistics and neurodivergent people, people with Asperger's syndrome were not considered "life unworthy of life."

The first people put into gas chambers by Nazis—before the 6 million Jews who would be slaughtered—were part of the T4 program that targeted disabled people. From 1939 to 1941, the T4 program gave a group of medical professionals the power to identify patients for "euthanasia" (mercy killing) who were deemed unfit. The first targets were children and adults with cerebral palsy, Down syndrome, and other disabilities from birth, followed by adults with chronic illnesses, mental-health issues, and non-German criminals. After lethal injection and starvation were decided to be inefficient, Nazis introduced the use of gas chambers disguised as showers. Often, families had no knowledge of what really happened to their institutionalized loved ones; they were told their family members had died of pneumonia and been cremated. Whole mental-health institutions were processed and killed, their friends and families given lies about outbreaks of illness. Disability studies scholars Sharon Snyder and David Mitchell work to share this part of disability history, the medical murder of an estimated 300,000 disabled people under the Nazi regime as part of work on a documentary called *Disposable Humanity*.[24]

We have to think of the larger picture here of how people are read and understood in society. This is not to say that we shouldn't work toward a cure for some disabling diseases or seek better medical interventions—but these interventions can't be only at the individual level, for that treatment has often led to awful things. We need a historical and political sensibility about disability issues. Some of "our" so-called charities have no understanding of or connection to our histories either. They downplay or outright ignore the context in which we exist, presenting disability as a matter of individual performance or function. There's been a ton written about the tactics of the Muscular Dystrophy Association, both by its former poster children and by other disability rights activists, but for the past two decades, the flash point for disabled critiques of disability charities (the rhetoric of their advertising and fundraising and even their goals) has focused on the organization Autism Speaks.

Autism Speaks—which disabled critics often delightfully render as "Autism $peaks," with a dollar sign like ke$ha—was founded in 2005 by Suzanne and Bob Wright, whose grandson had been diagnosed with autism. One of the group's commercials, called "I Am Autism," replicates many of the problems with their rhetoric and practice. Below, I quote the text from this commercial, which autistic self-advocates successfully campaigned to have withdrawn in 2007. This text was delivered with a sinister, gravelly voiceover and color-drained images of things like an empty swing swinging, without a child:

> I am autism . . . I work faster than pediatric AIDS, cancer, and diabetes combined. And if you're happily married, I will make sure that your marriage fails. Your money will fall into my hands, and I will bankrupt you for my own self-gain . . . I will make it virtually impossible for your family to easily attend a temple, birthday party, or public park without a struggle, without embarrassment, without pain. You have no cure for me. Your scientists don't have the resources, and I relish their desperation . . . I will plot to rob you of your children and your dreams. I will make sure that every day you wake up you will cry, wondering, *Who will take care of my child after I die?* And the truth is, I am still winning, and you are scared. And you should be.

This now-infamous fundraising commercial ends with replies from parents, siblings, and grandparents (notice: not autistic people themselves, who get no voice here) saying they are stronger than autism and united against it, ready to fight. The commercial first scares people, personifying autism as an evil force that steals children, comparing it to AIDS and cancer. This commercial frames autism as an outsider that ruins lives and families. But when you listen to autistic people talk, type, and write about autism, especially those supported and connected, you hear a different story. They talk about their brains as simply different, not scary, and about a world awash in misunderstanding for their expressions and behavior. To them, autism isn't something to fight against; it is fundamentally part of who they are.

We see in this commercial a deep ableism against autistic people, framing what autism is and what it does as a stigma. Up until very recently, Autism Speaks, a supposedly autism-serving charity, has had very few autistic members of their board of directors. Their funding allocations are also telling. They unfortunately put their money where their mouth is: they funnel funding in ways that reflect the same stigmatizing, harmful "awareness" rhetoric on display in their much-maligned commercial. In 2018, for instance, nearly half of Autism Speaks revenue was spent on awareness efforts and lobbying; 27 percent was spent on research, with 20 percent spent on fundraising and only 1 percent on family services (the category where spending would actually make the most difference for existing autistic people).[25] And as one Autistic Self-Advocacy Network (ASAN) member with username Savannah wrote in 2015: "As a lot of that research is genetic in nature, prevention means research into selective abortion of fetuses with markers for autism. Not only does this not help autistics of any age, it encourages the idea that it's better to not exist than to RISK being disabled (and in particular, autistic)." It's easy to see the actions of Autism Speaks as both connected to historical programs of human improvement and as part of wider charity messaging about disability where disability is always equated with grave tragedy.

In 2014, ASAN issued a joint statement against Autism Speaks's hateful rhetoric, with support from the National Council on Independent Living, Not Dead Yet, Little People of America, and Down Syndrome Uprising. It's heartening to see people speaking up and expressing/announcing their lived realities, though Autism Speaks still thrives financially as an organization and accounts for most top hits when you google anything about autism. Things are still made difficult by current approaches to autism, especially for those who are non-speaking or have additional intellectual disability.

Autism Speaks supports applied behavioral analysis (ABA), an approach to therapy that is currently contested by many autistic adults. Autism Speaks explains that it is an "evidence-based best practice treatment" based on the science of learning that uses positive reinforcement and the use of ABC (antecedent-behavior-consequence) sequences to "replace inappropriate behavior." According to Autism Speaks:

> ABA therapy includes many different techniques. All of these techniques focus on antecedents (what happens before a behavior occurs) and on consequences (what happens after the behavior). More than 20 studies have established that intensive and long-term therapy using ABA principles improves outcomes for many but not all children with autism. "Intensive" and "long term" refer to programs that provide 25 to 40 hours a week of therapy for 1 to 3 years. These studies show gains in intellectual functioning, language development, daily living skills and social functioning.

The use of ABA is extremely widespread; ABA is one of the few therapies for autistic children that is covered by most insurance, thanks in part to lobbying efforts by Autism Speaks—and the groups who develop and use ABA have been incredibly successful in both lobbying for the method to be covered and in increasing the ranks of professionals trained in ABA. Every year when I teach, I see a few students who are already interning at ABA clinics or may have plans to become BCBA (board certified behavior analysts) or registered behavior technicians (RBT). It can be a lucrative specialty compared to other disability service professions, and demand is growing, with each autistic child in what I would consider just a wild number of hours of therapy a week—some "professionals" recommend up to forty-plus hours of therapy for one to three years, even for very young children. One reason ABA is so appealing for some is that it amounts to many hours of paid childcare that a family might not be able to do without, and this treatment is covered by Medicaid and many insurance packages. This can't be said of other therapies that might be recommended for autistic people, such as occupational therapy and speech and language therapy or services for autistic adults around peer support, counseling, or technologies.

It's time to listen to autistic people about their ongoing opposition to ABA and other behavioral approaches to autism—illustrated only briefly here. In her blog, *Quiet Hands*, Julia Bascom writes about how different flaps and stims indicate different things and moods, and about having hands tied down or stifled in movement (to have quiet hands), about being encouraged to stop communicating when the communication is flapping: "If they see my hands, I'm not safe." She names ABA as abuse. She writes:

> They actually teach, in applied behavioral analysis, in special education teacher training, that the most important, the most basic, the most foundational thing is behavioral control. A kid's education can't begin until they're "table ready."

As Bascom writes, when she was asked to be "table ready," to have "quiet hands," she is being asked to put great effort into being still—which she experiences as being "deadened" and "reduced." She is being asked to "silence" her "most reliable way of gathering, processing, and expressing information." When ABA takes away stimming and flappy hands, it eliminates an important part of expression, communication, and sensory experience.

Bascom's critique of ABA therapy is widely shared. In the comments to the original post, people talk about the trauma they endured and the PTSD they developed in the context of ABA, and their stories show that stimming is crucial to emotional and sensory regulation for many people. It can be harder for some people to talk if they cannot stim. They stop speaking because of the overwhelm that stimming once helped them deal with by allowing them to self-regulate and process sensory information.

Finn Gardiner, a communication specialist with a master's in public administration, writes of a similar experience for the group blog *A Thinking Person's Guide to Autism*:

> My childhood and adolescence were steeped in the politics of shame. Family members and teachers reinforced the idea that I was intrinsically wrong just for existing. They may not have said it explicitly, but their actions were laden with this implication. I was told to have "quiet hands," to stop expressing myself in my natural, autistic way. I never knew why having loud hands was "wrong," just that it was. Meltdowns were seen as "behaviours" that needed to be extinguished, rather than expressions of frustration or overload whose cause should be investigated.

Gardiner here "rejects the politics of shame"—an important frame of reference, because ABA is set up precisely to erase autistic behaviors, positioning them as inherently negative. As he points out, much of his early life was steeped in shame, and he writes about how he has slowly learned instead to accept his autistic, racial, and gender identity. Many autistic people have written about how ABA rejects autistic communication, identity, and way of being, instead attempting to "normalize" behavior.

Another well-known autistic blogger, Amy Sequenzia, has written on the blog *Ollibean* about her experiences with ABA. Sequenzia, a nonspeaking autistic person who has worked with the Autism Women's and Nonbinary Network, writes in a very direct post entitled "Normalcy Is an Ableist Concept":

> ABA is damaging to Autistic people. It takes away true choice, it is bad for our self-esteem, it forces Autistic children to learn in ways that are not natural to the wiring of their brains, it interferes with how they process things . . . When parents of younger Autistic children say they are so proud and accepting of their kids, at the same time saying that they want the children to do things in a neurotypical way because society is not ready for the Autistic way of doing things, they are not being accepting, and they are putting a big burden on the child's back: change who you are or the world will hate you.

Sequenzia explains these ideas by sharing personal experiences and knowledge of communication and socialization.

Author Maxfield Sparrow, who writes the blog *Unstrange Mind* and has published two books on autism acceptance,[26] authored a post about ABA. He attempts to reconcile a long-standing tension between parents of autistic kids who believe ABA works because they see "progress" in their child and autistic adults who critique ABA as abusive. Sparrow writes:

The goal of therapy should be to help the child live a better, happier, more functional life. Taking away things like hand flapping or spinning is not done to help the child. It is done because the people around the child are uncomfortable with or embarrassed by those behaviors. But those are coping behaviors for the child. It is very important to question why a child engages in the behaviors they do. It is very wrong to seek to train away those behaviors without understanding that they are the child's means of self-regulation . . . You want to always remember a few cardinal rules: behavior is communication and/or a means of self-regulation. Communication is more important than speech. Human connection is more important than forced eye contact. Trust is easy to shatter and painfully difficult to re-build. It is more important for a child to be comfortable and functional than to "look normal."

Sparrow is also careful in this post to explain that some therapies that call themselves ABA aren't actually ABA but are called ABA in order to get covered under insurance and Medicaid in the United States.

ABA is touted in studies as effective, especially in the short term (though critics say its effects are overstated). But what critics ask us to look at is the long-term cost of these behavioral modifications and treatment. Critics (like the autistic adults quoted above) who highlight ABA as abusive—as producing PTSD, self-harm, anxiety, and depression among autistic people[27]—point to the entire goal of the therapy, which is on changing *behaviors* for the comfort and convenience of non-autistics rather than meeting the emotional and communicative needs of children receiving this therapy. This behavior-focused approach to therapy simply disregards—is designed to disregard—the feelings and cognitive changes of its subjects. Whether autistic people have internal lives and can think about their own thoughts was once seriously questioned by psychologists, and the idea that autistic people are purely rational like computers or have "ultra male" brains now tells us a lot more about the long-held biases against autistic people than it does about how autistic brains actually work.[28] As many autistic scholars have pointed out, non-autistic experts have mischaracterized autistic people's cognitive functions and experiences—often in ways that have dehumanized autistic people and made them subject to abuse. Indeed, people have questioned since autism was identified whether autistic people *even have feelings at all*, since they often do not communicate their feelings through neurotypical facial expression or other social norms. For the record, *autistic people have feelings*—whether or not they are intellectually disabled, whether or not they are cognitively different, whether or not they can speak, whether or not you can personally understand their facial expressions or behavior.

Mental content, feelings, any explanations that appeal to perceptions or other experiences of mind are secondary to behaviorism, if considered at all. Behaviorism will be familiar to many readers through Pavlov's famous dogs, said to have been trained to respond to a bell by drooling when Pavlov set up an expectation of food or treats.[29] This is classical conditioning—a pattern set up to produce certain behaviors. Behaviorism was in vogue in psychology for decades, and it's still a thriving branch of psychology. It is the central tenet of animal behavioral research, for example, where we can make guesses across cognitive difference, but behavior is what is most apparent and studied. It's also common in many human therapeutic approaches and research.[30] In all these areas, the focus of behavioral therapy is *modifying behavior.*

ABA therapy was "pioneered" by Ivar Lovaas, who used positive and negative reinforcement to shape the behavior of autistic children through discrete trials and intensive therapy. From the 1960s until his retirement, he was active at UCLA in developing ABA intervention. Obituaries and remembrances of him upon his death in 2010 tell a weird sort of tale. They paint a picture of a good man who was willing to work with and make progress on the most difficult cases (people). They talk about him as a luminary who founded a whole field, standardizing the method and using empirical behavioral analysis to decide on a scientific approach to reinforcements to change observable behavior. They talk about his charisma in the classroom and the many students he brought into the field. They talk about his success with the hardest cases (autistic children) he saw, remediating self-injurious behavior and teaching nonspeaking patients to verbalize. They emphasize the fact that ABA aimed to encourage positive behavior and eliminate the negative, and he is therefore painted as a success with just a few hiccups along the way, which led to interventions being made earlier (targeting preschool) and for longer periods (thirty to forty hours per week for one to three years).

In the obituaries, any controversies are conveyed by the barest of hints. But even during the 1960s, early in his career, there were questions about his methods, both within the research community and in the public arena. His use of aversives was highlighted in 1965 in an issue of *LIFE* in an article called "Screams, Slaps, and Love," which showed images of aversives—here, electric shocks—being used on crying autistic kids. Yet the obituaries and remembrances of him talk about him using positive reinforcement in "98 percent of cases" and championing the development of better methods of positive reinforcement.

But dehumanization of people who were not "normal" was at the core of all his work. One of Lovaas's famous statements is often quoted by critics of ABA who refer to its history and context. In an interview with *Psychology Today* in 1974, he said, "You see, you start pretty much from scratch when you work with an autistic child. You have a person in the physical sense—they have hair, a nose and a mouth—but they are not people in the psychological sense. One way to look at the job of helping autistic kids is to see it as a matter of constructing a person. You have the raw materials, but you have to build the person." The title of the interview is literally "Poet with a Cattle Prod," and even though the interview is somehow favorable, that title looks very different in light of current critiques and reckoning! And Lovaas's methods also constitute the basis for gay conversion therapy, now much more widely discredited (and outright banned in many states). Lovaas was a major player in the Feminine Boy Project (funded by the National Institutes of Mental Health, 1972–84), which employed many of his methods developed with ABA. Indeed, sometimes people refer to ABA as *autistic conversion therapy*. Both therapies have similar negative outcomes: high suicide rates, depression, lack of long-term efficacy, self-loathing, PTSD, and life made harder for lack of acceptance.

It's not incidental that queer-crip scholarship, a field where Nick Walker has coined the term "neuroqueer," has focused on the similarities of these methods, and the fields of disability studies and queer studies both roundly reject ABA and other forced behavior-modification programs as an appropriate approach. We now talk about homosexuality in terms of different preferences or orientations rather than as a disorder or pathology. The *Diagnostic and Statistical Manual* (DSM, the psychologist's bible) dropped homosexuality as a category of mental illness in 1974. Something similar has been happening recently with gender variance, with the DSM dropping gender identity disorder in 2013 and replacing it with the category of gender dysphoria. In other words, people are no longer diagnosed as gay or transgender as if it were medical deviance or disordered behavior. Autistic people are far more likely than allistic people (non-autistic people) to be transgender or nonbinary and to classify themselves as queer in some way. Connection between autism and gender nonconformity or variance makes a lot of sense. If autism is related to a person's sense of the social (autistic people are often said to have deficits on different measures of social behavior), and if gender is a cultural understanding that has many social components, why should we expect autistic people to have a sense of gender that aligns with cisgender and heterosexual norms?

The foundation of both types of conversion therapy, gay and autistic, is the attempt to make it uncomfortable or painful to be your natural self, in order to "normalize" your behavior for others' comfort or convenience. Former ABA therapist and blogger Anxious Advocate breaks down how behaviorism leads to compliance training in ABA:

> If a child has any behaviors that seem "repetitive" or "obsessive," or any that are just not understood by the neurotypical people around them, then behaviorists often try to change these things. Stimming is the most common example of this, but it could be anything. Walking on tip-toes, talking about the same topic too many times, using echolalia, having an intense interest, not making eye contact, covering ears, not playing with toys in a specific manner—anything that behaviorists deem "inappropriate" often becomes a goal the child has to work on changing.

ABA seeks to make autistic people seem less autistic. But when autistics are forced to mask and process in the world in ways that make neurotypical people happier, when they are trained to ignore the things that feel good and natural to them, it takes a huge cognitive and emotional toll, especially over time. Autistic people point to consequences of this type of masking: autistic meltdowns (sometimes called tantrums, these are in fact a manifestation of overstimulation—emotional outpourings that help them re-regulate) and autistic burnout (absolute exhaustion and even catatonia that can last for days or years from having to constantly perform as neurotypical). These are some of the results of having your natural behaviors bulldozed in the quest for normality.

Although modern ABA talks about moving away from aversives, the Judge Rotenberg Center in Massachusetts still uses electroshock aversives on autistic people, predominantly nonspeaking autistic people. Years ago, the United Nations classified the use of these types of electroshock devices in wartime as torture, but they are still authorized by the FDA and used on preexisting patients at the JRC.

Aversives tend to be used primarily on children who are marginalized in some additional way. Kerima Çevik at Intersected Disability has written powerfully about the activist work of mother Cheryl McCollins using the hashtag #AutisticWhileBlack. Çevik draws links between Cheryl McCollins and Mamie Till, mother of Emmett Till, who let the battered body of her innocent son be displayed to bear witness to the horrors of his death at the hands of racist cowards. In 2002, Cheryl's son André was rendered comatose by electroshocks at the JRC. She went to court to get surveillance footage of the events leading up to his coma. The footage shows André receiving thirty-one electroshocks. According to a 2012 story in *New York Magazine*, the devices used at the JRC are backpacks; the shock device "stays hidden inside, with wires extending from the backpack, running beneath their clothes, and attaching to electrodes strapped to their arms and legs. Staffers carry remote-control activators; when students display certain "targeted

behaviors"—like hitting, yelling, or trying to remove their electrodes—
an employee presses a button to deliver a two-second shock." McCollins
had worn the backpack for seven months. He was being shocked even
when his mother visited, when he was tied down—staffers shocked him
for tensing up his body. Upon witnessing this, Cheryl McCollins had to
act like she wasn't upset about it so that they wouldn't destroy the tape
as she worked to get it and ensure that others saw the abuses. There's
a huge campaign called #StopTheShock (use that hashtag on Twitter)
to call attention to the practice, supported by myriad disability-rights
groups, and protesters from ADAPT have staged rallies in the FDA
director's neighborhood to get him to ban the use of the devices. Autistic activist and scholar Lydia X. Z. Brown has compiled a long list of
activist work and writing against the JRC and its continued practices.[31]

At this point, you may be asking, "If not ABA, then what? What
are we supposed to do for kids who self-injure with their stims or who
seem to melt down for no reason, for those who can't communicate
what they want or need?" The answer is *listen to autistic people* about
what works best to help them adapt and thrive and communicate.
There is no one single answer, because there's no one way to be autistic. We can learn from autistic adults on how to adapt environments
and meet support needs. We can think about environmental changes:
getting rid of fluorescent and harsh lights, having non-itchy clothing,
having tools like weighted blankets and fidget devices, making sure the
buzz of things around us stays low, providing quiet areas and times,
offering clear schedules and directions, allowing people to stim in ways
that don't harm anyone. Asking what works for people and following
those directions can be key.

It's easier than ever to listen to the experiences of autistic adults,
and there is work out there accessible to parents of newly diagnosed

children or to newly diagnosed adults. We now have books like *The Real Experts*, edited by Michelle Sutton. There are resources for parents from the Autistic Self-Advocacy Network and the Autistic Women's and Nonbinary Network. There's a whole autistic-run press called Autonomous Press that only puts out work by neurodivergent writers, some of which is aimed at the community and some of which is aimed at newly diagnosed adults and parents looking for answers. Robyn Steward's book *The Autism Friendly Guide to Periods* breaks down all the stressors of menstruation and walks people through their body's changes and what to expect during puberty—which is an excellent resource whether or not a preteen is autistic. ASAN has great lists of books and guides. There are so many more autistic-led resources out there right now on social media, too. I've mentioned several blogs and social-media hashtags already. I have become a huge fan of the TikTok account that details what to expect when you go into different places (so you don't get overwhelmed when you visit): quick guides to post offices, Aldi, and other businesses for those who might be more comfortable going if they know what to expect.[32]

As someone who wants to "fight technoableism," I am thrilled to see autistic scholars creating and promoting technologies and therapies that cohere with autistic experience and critiquing tech projects aimed at autism that don't line up with autistic experience. At an event on technology and disability that I organized,[33] we ran a panel about autism tech. Panel members were one non-autistic researcher (my disability studies colleague Carolyn Shivers) and four autistic scholars (Rua Williams, who is with Human-Computer Interaction at Purdue; Finn Gardiner, who does policy work; Elizabeth McLain, who is a musicologist and one of my good friends; and Reem Awni, a grad student in the Disability Alliance who did software engineering in Silicon Valley before coming to Virginia Tech to practice engineering design for social justice). Philosopher of technology Damien P. Williams, who works on bias in AI and biotechnology, moderated the panel and led with the right questions for this conversation. The discussion centered autistic experiences with environments and technologies. It didn't cover the neurotypical things you'd find in any special ed or human development textbook. What the true experts had to say was startlingly different—and refreshing.

In the session, the group spent a surprising amount of time talking about Pokémon and Dungeons & Dragons as autistic technologies. Elizabeth McLain brought these up as examples of cultural technologies as she explains her work in disability culture:

> I don't know if D&D is actually an autistic technology, but there are a lot of dungeon masters who are autistic, and we create our own autistic technologies in the space. We know how to Pokémon. . . . It's funny when you let us use our technologies or let us take other technologies and encrypt them and make them autistic technologies. We can socialize, we can connect, we empathize, we do all of these things that the stereotypes say that we don't do. We do them differently.

McLain suggested that these types of cultural technologies, which are so radically different from the scientific approaches to autism research or the digital interventions into social skills, invite us to think differently about technologies for autism:

My advice would be to find these autistic technologies, learn about how autistics use them, and engage them on their own terms. When we Pokémon, we collect them all and we will tell you all of the stats . . . This is not us being in our own little world, it's sharing a world that gives us joy with you. It's actually sharing a deep part of our soul with you. You need to get on board with how our culture works and how our technologies work.

Finn Gardiner followed up on this, adding:

. . . video games such as Pokémon . . . are often more effective as models of social behavior than the classic ABA style, "here is how we're going to make you look normal, be indistinguishable from your peers" . . . When we get into our intense interests . . . it's a way to express our enthusiasm for the world and the people within it. Pokémon is a particularly good example because the characters tend to talk about the importance of friendship, the importance of listening to people, and the importance of the partnership between you and your Pokémon.

(He observed: "Some of the aspects of the game, for example, like whenever somebody gives you eye contact, that starts a battle, that might resonate with a lot of autistic people.")

This led into a broader discussion of games and digital technologies that work well for autistic people. Gardiner mentioned the Sims as another game that autistic people like, and later Awni mentioned Discord as a social platform "easily adapted by autistic users to facilitate autistic-styled communication due to its flexibility both with custom emotes and for purpose-centric server organization." Gardiner pointed to the importance of letting autistic people lead tech conversations with what they want and how they want to do things. This doesn't mean jumping in to teach autistic people to play games in the way allistic people do but rather letting the "nothing about us without us" lesson of the disability rights movement carry into this space: to learn from autistic people how to use spaces autistically.

The panelists as a group pushed against autism technologies or interventions (like ABA) that don't have the support of autistic people, and for the need to recognize autistic expertise and create power for autistic people in conversations about tech. And they pointed out that, as with Dungeons & Dragons, autism tech is not always what people expect. For example, Awni said: "My current favorite autism technology isn't normally considered one. It's actually my digital typewriter with no Wi-Fi connection, and therefore no distractions."

They discussed the problems autistic researchers face within the deeply ableist system of academia. Awni recalled presenting at a neural engineering conference where "the poster right next to me purported to be trying to develop an intervention to cure autistic people, but specifically focused on autistic model rats." Awni had asked the researcher how they had replicated autism in rats. According to the presenter, the rats "had a high propensity for fear." Awni said, "When they imagine somebody [autistic], genuinely it is coming from a paternalistic place of complete and total ignorance." The panelists then had a whole aside about the supposedly autistic mice being a very bad approximation for anything autistic, calling the models laughable, ridiculous, and based on a poor interpretation of autistic behavior.

Human-development scholar Carolyn Shivers explained that one of the ways they produce "autistic" mice is by cutting their corpus callosum—which, of course, has nothing to do with the mechanism by which autism occurs in people. When autism is studied in mouse models like this, misunderstanding arises in serious ways—both among researchers, like the poster presenter next to Awni, and among the general public who are looking for information for their loved ones. I see a similar thing happen regularly in my cancer groups. People will get very excited about "in vitro" studies of different compounds killing cancer cells and then project that into all sorts of things they could try at home, not understanding that "in vitro" is in a petri dish (and "in vivo" would be in an animal). Not that the research isn't exciting, but it's not very far along, reliable, or useful for human application at that point. Shivers explains how much of autism tech intervention research re-emphasizes pointless norms of communication or creates very strict parameters of norms for behavior to "teach" autistics to appear less autistic in different scripted scenarios—through games with robots, virtual-reality simulations, and video modeling. Panelist Rua Williams picked back up on this: "It's like on the one hand, you'll say autistic people are broken because they have scripted social responses, but we gave them normative scripted social responses." Williams was surprised they do not recognize this irony. None of these tools, supposedly created for autistic people, was anything our panelists saw as exciting autistic tech. And panelists agreed that the autistic tech they do like should not be made into therapy. Things should be allowed to be joyful and fun and autistic without turning them into exercises for social scripting and normative behavior patterns.

Williams has a human-computer interaction (HCI) lab that uses "counterventions" (their term)—counter-interventions that let disabled participants involved in participatory research intervene in correction of researchers' assumptions and change the direction of tech projects to create disability-led (and in the case of their most recent research, specifically autistic-led) technologies. Williams explained for the panel that 90 percent of technological interventions for autism have been targeted toward social skills—digital experiences meant to "train normative social interaction." They explained that this type of intervention leads to high levels of suicidality among autistic people: "When the vast majority of our intervention, and in particular, our technological intervention for autism is targeted at social skills, we're basically producing the suicidality of autistic children." Williams envisions approaches that flip the tables and center autistic norms and culture in how we approach making technologies. They are putting forward proposals where autistic kids direct designers about what works for them (centering what the kids want to work on and goals they have, not the ones researchers have).

McLain pointed out that most studies characterizing autism are of white eight-to-twelve-year-old boys. She also noted that the stereotypes of autistic people as robotic are often drawn from examples of *traumatized* autistic people. This is because many autistic people have been traumatized—if not through ABA therapy, then through living in a world that doesn't regard them as fully human, one not adapted to autistic needs and embodiment. As McLain put it, "We are not robots. We are just living in a constant state of trauma." She talked about the joy of being together with other autistic people, where the norms shift to autistic norms and nobody needs to mask. Remember I related in Chapter 4 about the joy of going to the amputee convention, and how restful and amazing it was for me to not be stared at and to be able to dance with people moving in so many different ways? These small spaces adapted to disabled bodies and minds, filled with other disabled people who understand our experiences, are such an important refuge from a hostile world. McLain explained, "Our experience of autism is very different than an outsider is going to describe it." This is why *listening to autistic people*, whether they talk or type or point at pictures, is all the more important to autistic flourishing, and to good autistic tech.

McLain is my friend and colleague, which means that we get into lots of disabled trouble together. She's a more recent arrival to the Disability Alliance and Caucus. Last year, it felt like the DAC was coming full circle when Phoenix and Elizabeth got to meet at an online DAC graduation event, where they were both graduation speakers (alongside staff member Annabelle Fuselier; we like to have a whole community of speakers).

I've been thinking a lot lately about the uneven lines between impairment, difference, and disability. I've gotten used to my new chemobrain and adapted to what was an impairment. My experiences of memory and time and finding the right words are all changed. I used to wake up every morning and know what day it was and what was on the calendar for that day before I even opened my eyes. That's gone. Now I consult my smartphone. It's not a big deal. For a while it was, though. I forget meetings a lot more than most—which is why having good people around matters. Elizabeth texts me if we're supposed to be at a meeting together (and vice versa), and then we both make it where we are supposed to be.[34] It took me too long to figure out this hack (from another DAC member, Liz, who has ADHD), but I set alarms ahead of all the things I need to do regularly on weekdays: dropping off and picking up my kids, going to teach my own classes, and so on. Impairments—lower functional differences—can exist without being or staying a problem.

Other impairments can pester for life. I don't think I'll ever fully adjust to my tinnitus, this constant high-pitched screaming in me that no one else can hear. But here, too, I take some joy in community. One of my favorite memes[35] of all time is from Reddit's r/ADHDmemes area. It is a picture of three people crouching around and petting the same dog. The people petting the dog are labeled: ppl w tinnitus, ppl with ADHD, and ppl with autism, and the dog, a sweet golden retriever, is labeled "dog"; in the middle of the image, where the circles in a Venn diagram would overlap, is the phrase "the world is full of high-pitched sounds nobody else can hear." We can share experiences and not be the same, and we can laugh about it sometimes.

CHAPTER 6

Accessible Futures

THE FUTURE IS DISABLED

Technofuturists want to believe that in the future utopia there will be no disability. But despite eugenic and transhumanist imagination about perfecting human beings, about life extension, about regenerative medicine, about neural implants, and the merging of "man" and machine, the fact is that in the future, we can expect more disability, not less. We should expect both more ways to be disabled and more people existing with disabilities. *The future is disabled.* This is a tagline of the Disability Visibility Project and the title of a new book of poetry by Leah Lakshmi Piepzna-Samarasinha. We can mean this politically, as they do. We can also mean it quite literally. When I say the future is disabled, I am stating a fact along every axis of consideration I can imagine.

The future is disabled for each individual. We can never have too many reminders that anyone who lives long enough can expect disability eventually; disability is a very normal and predictable part of the human experience. There are significant social differences between those who acquire disability in old age and those of us who arrive there much younger or are born disabled. We are often pitted against one another for what is seen as limited service/funding/care, and we experience different social biases and expectations. Yet we are all part of the larger disability community, whether we like it or not. Sometimes older people don't want to admit this; they want to distance themselves from disability as a category. Sometimes disabled youngsters don't want to see it, bristling at being lumped in with older people. But we are all in this together. It's important we treat disability from aging as respectfully and inclusively as we do the other types of disabilities I've discussed in this book. Enough with the jokes about the disabilities that come with aging—jokes about Depends, hearing aids, orthopedic shoes, and mobility aids. These are all disability technologies that are also used by young people, and they aren't good symbols for growing old. We need to be making better plans to be and become disabled, now and in the future—and as we design places where people can age in place, we need to not leave out young people. This is especially true in the era of Covid, which is disabling young people in record numbers with all the various impacts of long Covid (with estimates ranging from 8 to 25 percent of people who have had Covid); the disabled future is coming to pass now, and we need to create inclusive and accessible environments for all kinds and ages of disabled people to deal with it.

The future is disabled for humanity writ large. As a species, we face environmental disaster, pollution, and climate change. Environmental racism has already wrought disproportionate damage to human health with our decisions about where to site environmental hazards like interstates, gas pipelines, landfills, and so on, and this will only get worse as the number of healthy, habitable locations is reduced by climate change. Pollution is increasing rates of environmentally produced disease and disability—higher levels and lower onset ages of different types of cancers, as well as rising rates of asthma, chemical sensitivities, and autoimmune disabilities, some of which can come from smog and conditions of poor air quality. Think about all the people along the West Coast who breathed in toxic smoke from wildfires for weeks on end in 2021: these kinds of environmental health hazards are only going to increase. We've seen successful campaigns against some types of environmental health and work hazards: with our knowledge and observations about lead (contributing to mental disability and lead poisoning) prompting policy changes to materials used in gas products, paints, and more. Pollution and environmental damage are not inevitable, but we will be managing the disabling outcomes for a long time of things that have already happened.

The future is disabled for the planet itself. Sunaura Taylor, a fellow disabled scholar who is an animal and environmental activist, writes powerfully of the "disabled ecologies" that constitute the landscapes we have impaired. Her case study is the Superfund site in Tucson, Arizona, which contaminated local groundwater and, forty years later, is still affecting the land and communities nearby. She thinks disabled people have important insight into how to live and age and exist with disabled ecologies. She reminds us that we can't just get rid of our land, our environment. We have to learn how to live in a world we have disabled.

I am writing this book from where I live near Virginia Tech in Blacksburg, on the traditional lands of Tutelo and Monacan people.[36] Blacksburg is a bucolic area, with cow pastures and forests and close access to the Appalachian Trail and the New River. But I live less than ten miles from where the Mountain Valley Pipeline is being built. For many years, people have been fighting hard to stop its construction, but giant equipment has moved in, people's farmland has been ruined, and locals have been bought out and displaced by MVP—all as environmentalists and residents have continually pointed out the cost and risk in erosion, leaks, and explosions. We have locally famous "tree sitters"—people who have slowed down and blocked construction by staying up in tree stands for hundreds of days at a time. I have fierce friends who have supplied and sung to and visited the tree sitters. (I have been invited, but friends were frank with me about the terrain as we talked about what it would take for me to get there and back. Steep slopes are hard.) I found out my friend and colleague Emily Satterwhite, whose work is in Appalachian Studies, had locked herself to construction equipment when I was at the Library of Congress for a "Decolonizing Mars" Unconference, which I will discuss later in this chapter. I also live just twelve minutes down the road from the Radford Army Ammunition Plant (RAAP), which manufactures propellants, explosives, weapons systems, and artillery. It uses open burn for some of its waste disposal and its permitted emissions make it the biggest polluter in the Commonwealth of Virginia.[37] The tranquility of rural and town life in and around Blacksburg is but a surface waiting for a scratch to reveal the ways in which we are being invisibly poisoned by greedy developers and militaristic bureaucrats.

The future is disabled, through changing weather patterns and natural disasters. Even if specific locations and corporations are checked, the oncoming climate catastrophe will affect us all eventually. Changes in climate bring changes in weather, and we will see more natural disasters and the disability that comes from that trauma. Climate change has already brought new geographies of tick-borne and mosquito-borne disease. This will intensify, and we will see disease patterns and regions shift as a result—often to places that are ill prepared or ill equipped for vital treatment or effective communication about public health. New diseases will emerge as animals are driven out of their habitats by climate change and come into contact with more people, making the jump from animal to human more likely. Some of these diseases will have a long and unanticipated tail.

The future is disabled, through emerging and new diseases. Like other post-viral diseases (which those in the disability community—especially those with myalgic encephalomyelitis, which used to be called chronic fatigue syndrome, and POTS, have been talking about since mid-2020), long Covid can cause difficult-to-diagnose, difficult-to-treat conditions. We are already seeing an onslaught of long-term vascular, neurological, pulmonary, and cardiac-related disability from Covid infection. We remember polio mostly for the long-term disability it produced (iron lungs, paralysis), not the short and often mild initial infection, which simply caused fever and stiffness. Viruses are weird and change us for the long term—often in ways that initial mild infections do not prime us to expect. As we get new and mutated viruses, we get new forms of disability. Indeed, disability activists have been calling Covid a mass disabling event.[38]

The future is disabled, cosmically. Even with hopeful futures like space travel and adventure, we can expect the production of disability. Space is already disabling for humans. Just as the built environment on Earth is not suited for disabled bodies, space as an environment is not suited to *any* human bodies. Every astronaut comes back from the low gravity of space with damage to their bones and eyes—and the longer they are off Earth's surface, the worse the damage. Some things can be restored over time, but some changes are long-lasting. These realities are absent from futurist writing about technology, which is framed as simply magicking away the disabling effects of space travel.

This is why technofuturists' discussions of "The End of Disability" are so silly. Disability isn't ending; we're going to see *more* and *newer* forms of disability in the future. This doesn't mean that all medical projects aimed at treating disease and disability are unpromising. I certainly hope we will come up with better diagnosis and treatment for conditions like asthma and Lyme disease. But disability is multi-faceted and a much larger category than can be covered by any simple case. *No nondisabled person without experiential knowledge of disability and engagement with the disability community should be making claims or decisions about the future of disability and disabled people.* We need to prepare for the disabled future, becoming more comfortable with other people's disabilities, accepting the fact that we ourselves will eventually be disabled (if we aren't already), learning to recognize and root out ableism—these are all moves toward building a better future for everyone.

Because the future is disabled, working toward and planning for a future without disability, without disabled people, is a *ridiculous prospect*.

How do we prepare for this disabled future?

In 1979, moral philosopher Alasdair MacIntyre (then fifty, now deep into his nineties) published my favorite paper for bioethics discussion. The paper, called "Seven Traits for the Future," was published as part of a *Hastings Center Report*[39] in a special issue on "Designing Our Descendants." MacIntyre writes about seven traits that we should engineer into our children, but flips it: for any of the traits we would value, that value itself is poised against altering our descendants. It's a cool philosophical move! But here, I only want to talk about the first trait he discusses: the ability to live with uncertainty, which is certainly something we would want our descendants prepared for. He writes:

> To an unspecifiable degree the future is unpredictable. This unpredictability immediately raises difficult questions for any project of designing traits for the future, since no trait can be thought of as being good or bad independently of that environment in which it flourishes, toward which it will organize human responses, and from which it will derive material for its tasks. Thus if we cannot predict the environment, we have an initial problem in selecting desirable traits.

I love that "no trait can be thought of as being good or bad independently" of environment, response, and task. Reminds me of the social model of disability and of disability tech itself, even though that's not what he was considering in the terms that I usually do. He ends the paper with the following observation: "If we conclude that the project of designing our descendants would, if successful, result in descendants who would reject that project, then it would clearly be better never to embark on our project at all." We can't design well for that which is unpredictable. But we can design more accessibly and can invite futures that enable more of us to live and thrive, by investing in infrastructure and programs and networks of care that "make sure we all make it" (to borrow from poet Leah Lakshmi Piepzna-Samarasinha, who puts everything better than this non-poet could).

The unknowability of the future is also addressed by science-fiction writer Ursula K. LeGuin in *The Left Hand of Darkness*. LeGuin describes a conversation between our main character, Genry, an alien on a strange world, and Faxe, a foreteller or weaver for the community.

"The unknown," said Faxe's soft voice in the forest, "the unforetold, the unproven, that is what life is based on. Ignorance is the ground of thought. Unproof is the ground of action. If it were proven that there is no God there would be no religion. No Handdara, no Yomesh, no hearthgods, nothing. But also if it were proven that there is a God, there would be no religion.... Tell me, Genry, what is known? What is sure, unpredictable, inevitable—the one certain thing you know concerning your future, and mine?"

"That we shall die."

"Yes, There's really only one question that can be answered, Genry, and we already know the answer.... The only thing that makes life possible is permanent, intolerable uncertainty: not knowing what comes next."

I thought about this section of LeGuin's work through much of my cancer treatment—both during my initial year of chemo and amputation and during both of my incredibly terrifying recurrences.[40] I felt like everything had been yanked from my control—the future as I knew it was gone, uncertain, and I was unable to plan or prepare for it. With a diagnosis like mine, even day-to-day things, small controls over one's life, slip out of your hands. That transition, before adaptation and reconciliation, is so hard.

When I was diagnosed, I was a normalish thirty-year-old at the beginning of a career. I had a PhD, an enviable academic gig, and I was happily married with two kids. I even had an enviable work-life balance, working only half-time so I could spend more time with my kids while they were little. My kids turned three and five during my year of treatment, and I missed so much time with them while I was hospitalized.[41] I worried I would die without my sweet, smart, sassy three-year-old having any memory of me, and with many of my five-year-old's memories being of the worst time in our life. And my cancer is predominantly a pediatric cancer, so the others I heard about through cancer groups going through the same brutal treatment as me were mostly children.[42] I know some things about intolerable uncertainty. Indeed, I still revisit a sliver of that uncertainty every three months, when I have monitoring scans—a regularly recurring trauma that I have to wait for and worry over. (My cancer has recurred twice.) This is not to paint myself as a tragic figure—it sucks, but many people know and experience the same thing. And we had so much help over the course of that year from family and friends.[43]

I was lucky to have the right support to keep my kids in a routine and in a familiar place; I know people in my cancer community who temporarily lost custody of their children or had to rely on people who were less than trustworthy. Some lacked partners and friends who could drive them, sit with them, pick up prescriptions, handle some of the paperwork of being sick. Everything was as good as it could be for me, which is to say that it was still awful, but wasn't awful for lack of support. It was awful because chemo sucks. My mouth was full of sores. (I still have permanent discolored gray spots in my mouth and on my lips that started during that time.) I spent a month in the middle of treatment with a yeast infection in my mouth—no immune system to step up to the challenge. My tinnitus blared. One chemo drug made my palms and soles swell and get sore, so that was a recurring part of the treatment cycle. By my fourth cycle (of eight), I would get the shakes a few days into the administration of one of the chemos and couldn't hold a glass without dropping it. I fell twice in the hospital due to shaky hands on my walker—the second time I fell, I was trying to explain how I had fallen the first time. After that, I was not allowed to go to the bathroom by myself and had to mash the Call button and wait—a wait made more difficult by the fluids they kept pumping through me.

The future is uncertain, and disability makes it feel more so. Not that disability makes things more uncertain, but it makes you see that before, the predictable future of permanent health and safety you had comfortably imagined had, in fact, been an illusion. I am sure there are other experiences in human life—displacement, natural disaster, crises of all sorts—that bring uncertainty into focus the way disability does, but this is how it happened for me.

Disabled people know a lot about uncertainty—if it isn't cancer-land and scanlife, it's a surprise flare and varying symptoms of auto-immune conditions; it's finding out there's a car parked too close to your lift-equipped van; it's a disruption to a plan that sends you toward a meltdown; it's that everything is itchy and you have to leave; it's a hearing-aid battery dead; it's that the elevator is out and the event will proceed without you; it's that no one has planned how you'll be evacuated in an emergency (they didn't plan for you at all).

When we ask for the ability to live with uncertainty, we are asking to learn "the fine art" of being disabled. To grapple with the uncertain future, we would be well served to listen to disabled experts. The experiences and authentic stories of disabled people give us a lot of insight into our technological futures.

- Disabled people have expertise in navigating worlds not built for us—worlds that are often actively hostile to us.
- Disabled people know what it's like to deal with the unplanned, and to be unplanned-for.
- Disabled people know about systems of maintenance, repair, and upkeep; we know that you can never depend on "the mar-

ket" to bring down prices. This has never happened for much of what we use, and developing good things (desired tech, drugs, therapy) does no good if people don't have access to them.

- Disabled communities know a lot about the work of care, and about how to care for one another under hostile conditions. We also know the cost and consequence of care being unavailable, negotiable, or restricted.
- Disabled people know what it's like to exist and persist, even in the wake of attempted elimination, trauma, and grief.

It's not just disabled people who understand uncertainty and injustice. Other groups also understand uncertainty intimately, and disabled people come from every other intersection of identity possible. I am often moved by Afrofuturism and contemporary Indigenous narratives of futures that speak to how to live from the place we are in now.

One group of disabled artists-activists who seeks to capture both the uncertainty of our future and the importance of nevertheless moving toward that future with hope and intention is Sins Invalid, an artist performance collective out of San Francisco that leads the disability justice movement. Over the years, Sins Invalid has been led by and included people like Patty Berne (co-founder and executive and artistic director), Leroy Moore Jr. (Krip Hop Nation co-founder), Leah Lakshmi Piepzna-Samarasinha, Nomy Lamm, Eli Clare, Cyrée Jarelle Johnson, Lateef McLeod, Stacey Milbern, Laura Hershey, and many more luminaries involved with disability justice and the arts. They have put on different performances and shows that center the experiences of multiply marginalized community members— disabled trans and queer people of color. The group describes their most recent show, 2021's *We Love Like Barnacles: Crip Lives in Climate Chaos*, this way:

> Holding space for love, mourning, and community healing in pandemic times, Sins Invalid brings forth a performance that centers our communities in the throes of climate chaos and our agonized planet. From the storms battering our shores to the raging fires threatening our homes, the social, political, and economic disparities faced by disabled queer and trans people of color put our communities at the frontlines of ecological disaster.
>
> With the pandemic laying bare the inequalities and injustices that continue to exist in our current oppressive systems, now more than ever is the time to listen to crip wisdom in the age of climate chaos. Join us as we speak our truths on stage.

In addition to their performance work, Sins Invalid offers educational materials and sessions, including the cornerstone documents of disability justice as a philosophy and movement designed to meet our current circumstances. They have articulated ten principles of disability justice: Intersectionality, Leadership of those most affected, Anti-capitalist politics, Cross-movement solidarity, Recognizing wholeness, Sustainability, Cross-disability solidarity, Interdependence, Collective Access, and Collective Liberation.[44] All of these are ways of working in community to live on an increasingly disabled planet.

Other artistic movements about disability and the future echo this long-standing San Francisco group in different ways. Deaf-and-disabled-led literary journal *The Deaf Poets Society* asked us to dream in 2017 with their #CripsInSpace special issue. Guest edited by Alice

Wong and Sam de Leve, this issue was announced with a video of de Leve showing us how they are specially suited for space, for as wheelchair users, they were already trained to push off of kitchen counters and walls to get where they wanted to go. They also pointed out that most kids could dream of being astronauts, but disabled people are usually given fewer options, even early in life. And so they asked us to dream, write, and create art: the issue features short stories, prose, and poetry in which people think about how they are better suited for going to the stars. Marlena Chertock writes:

> *I would go to Mars if I wasn't too short*
> *for NASA's height restrictions.*
> *I'd tell them you can fit more short people*
> *into a rocket. Don't worry*
> *about my bone deterioration rate,*
> *I had arthritis at 13. Walked like an old lady*
> *at 20. It'd be nice to float*
> *and give my bones a break.*[45]

Others have also considered disabled space travel and disabled futures. In 2018, Sheri Wells-Jensen, blind linguist, banjo player, and my Facebook friend, made "The Case for Disabled Astronauts" in *Scientific American*. She wrote about how useful it would be to have a totally blind crew member aboard. Spacesuits would need to be better designed to transmit tactile information, but a blind astronaut would be unaffected by dim or failed lighting or vision loss from smoke, able to respond unimpeded, unclouded, to such an emergency—Wells-Jensen refers to a problem on the *Mir* where they couldn't find the fire extinguisher when the lights went out.

Two discussions at the Library of Congress about uncertain space futures took place in 2018, first an "un-conference" called Decolonizing Mars and a few months later a series of panel discussions and performances on becoming interplanetary. These events, organized by astronomer Lucianne Walkowicz (who has since founded the Just Space Alliance), fostered conversations from a wide variety of perspectives on how our narratives about space center "The Right Stuff" (to borrow the title of Tom Wolfe's novel) in ways that are sometimes problematic when it comes to recruiting, dreaming, and planning for space. The "stuff" that is taken to be "right" is usually privileged, masculine, from dominant cultures, and *extremely* abled (there are stringent physical "fitness" requirements for astronauts). Recruiting for space has always held up certain bodies as better than others, in ways that don't at all map onto what might actually work best. During the Decolonizing Mars event, as we sat in a smaller group discussion circle, I learned that short women with larger thighs do better at not passing out when they pull high numbers of g's as fighter pilots; their brains are closer to their hearts, so the additional blood flow helps them remain conscious, and their larger butts/thighs seem to absorb some impact. Yet typically, the "best" fighter pilot looks like Val Kilmer as Iceman in *Top Gun*.

Later I offered the example of the Gallaudet Eleven—eleven Deaf men recruited from Gallaudet University in the 1950s and '60s for a NASA study of motion sickness. They went through astronaut training and many different tests. Congenitally deaf people don't get motion sickness, and NASA simply wanted to know how nondisabled astronauts could avoid motion sickness too. The Deaf men were never considered for astronaut candidacy, however, despite their ability to avoid motion sickness. Other participants[46] highlighted the ways in which our space rhetoric perpetuates narrative structures that have done a lot of harm—ideas about frontiers, claiming planets and territory, mining and extraction from other planets, and colonization. As they pointed out, the continued use of these terms restricts how we imagine space, framing it simply as a continuation of colonization and capitalism—the very ways of thinking about space, ownership, and land that are so deeply disabling the Earth. As Black astrophysicists shared stories about the racism they encounter in their field (a field they love but doesn't love them),[47] we discussed how meaningful voices and stakeholders on these topics are often excluded. And later, my nerdiest dreams collided with my coolest hopes as Sammus the Rapper (aka Dr. Enongo Lumumba-Kasongo) performed for us in the Library of Congress.[48]

Sheri Wells-Jensen has now been on a zero-g parabolic flight to feel what it would be like to be in space: she and others put her case for disabled astronauts out into the world, and onto the right desks—and she became part of the first flight of Mission: AstroAccess. The goal of AstroAccess is to include disabled people in space exploration. Their first mission flew with twelve disabled "ambassadors" aboard in 2021, and they flew again in late 2022. To me, this disabled zero-g flight was huge news, just as Stephen Hawking's similar zero-g flight had been in 2007. However, the AstroAccess flight made less of a public splash; I only saw it reported because I follow disability-specific news.

In fact, it's a big time for disabled people in space. In my pediatric cancer groups, people were celebrating the SpaceX orbital flight of Hayley Arceneaux, who has an internal prosthetic knee (fake bones). The European Space Agency also announced its recruiting program for "Parastronauts"—a "feasibility study" for people with short stature (little people and others below a height threshold who would have once been barred from consideration) and lower-limb "deficiencies" (below-the-knee amputees, as well as people with foot-related issues like clubfoot or injuries to the ankle or foot). I can honestly not think of any reason why inclusion wouldn't be feasible. Just an amputee here imagining environments where walking is not the primary mode of locomotion!

The thing about space flight, about space stations, and the type of exploratory travel we are talking about in space, is that it's *all* uncertain; we don't know what skills might be needed. (This is true on Earth, too, just easier to imagine off-surface.) And all the necessary space infrastructure—any aircraft, spacecraft, or station—is something we build. (We could certainly be building regular airplanes to be more accessible to disabled people already: wheelchair users are particularly degraded, restricted, forgotten, and excluded with our current airplane setups.) We already know retrofitting sucks. Why not build things to be as inclusive as possible now instead of trying to fix them later? Finally, since we are going into an environment that we were not brought up in, it doesn't matter whether astronauts are nondisabled: again, we are *all* disabled in space. Our environmental niches are all on Earth, and our capabilities are all Earth-related. Disabled people don't have the same disadvantages in space that they may have here on Earth—especially if we work to avoid creating or re-creating disadvantages in how we build and plan for space.

My disabled friends can imagine ways that we would be well suited for space or space for us; we can all give different reasons why either our bodies would feel better in space (less gravity weighing our pain on us) or about why our bodies would be superior for space flight or travel. My friend Mallory is the cleverest here, because she's well adapted for pooping in space. In case you don't know, it's very difficult to poop in space—both in terms of physical properties and engineering-wise. Astronauts have to train to use specialized toilets (there is a whole toilet-engineering team with each space agency), and the toilets are

finicky and have a history of breaking. Because pooping is so complicated, Mallory has suggested that NASA should only be recruiting
people with ostomies—people who have openings in their abdomens
(called stomas) to excrete waste using ostomy bags. All the engineering
and work that currently goes into space toilets is only necessary because
no one has an ostomy![49]

I'm puzzled about why we aren't actively recruiting *for* some types
of disability here. Sheri Wells-Jensen has already given us the case for
the edge blind people would have on crews, and Sam de Leve as part of
#CripsInSpace discussed the edge manual wheelchair users would have
in moving in space. The Gallaudet Eleven were considered superior,
studied for that reason![50] I got to lead a colleague's class at one point—a
fun class about planning for Mars, not specific to disability—where we
ended up talking about how people who have experienced some types
of mental illness might be better suited in some ways to monitor themselves and others around them for certain emotional and physiological
responses to space. They might also help come up with ways to manage conditions like seasonal depression, which could be a huge concern
if we traveled farther away from the sun. We already know what our
slight tilt of the Earth means for mental health in the far north and far
south of our globe when it comes to higher rates of suicide and depression, and we should be attuned to this in how we plan for space too.

We need to be wary of technoableism—technology development
and marketing that makes it seem like disability is a big, bad thing that
needs to be downplayed or eliminated. Most of our supposed experts
about disability are nondisabled people, who don't know what it's like
to be the object of ableism, of design made at you rather than for you,
of future imaginings that snuff you out of existence, of scrutiny around
every one of your choices, your behavior, and your being. This is why

we need to look to intersectional, cross-disability communities for expertise—creative visions of a future that cuts no one out. We need to make the world more hospitable to more ways of being and existence—not just by heeding disabled expertise but by loosening our ideas about what "the right stuff" is and by insisting there is no wrong stuff. We should be actively anticipating all the stuff—and planning in a vein to mitigate damage already done in our disabled ecologies and prevent making our planet even more uninhabitable for some or for all.

HEARING STORIES FROM OTHERS' perspectives is fundamental to preparing for uncertain futures—and to have imagination about how the world could be, and could be different. Such stories should galvanize how we consider the future.

When my kids were in elementary school, I would show up for different school activities and, for a while, read weekly in my kids' classrooms. I'd bring some disability-centered kid lit, but also just fun books too.[51] As an amputee, I was usually going in with a prosthetic leg and a cane (and sometimes with crutches). I wore hearing aids too. I love sharing stories with different characters, historical and fictional; kids are much chiller than adults about differences and ask better questions. They aren't attuned to the same pity, shame, or concern that adults have been conditioned to be by our culturally dominant stories about disability. As soon as I told the kids that I wasn't in pain (sometimes kids would be concerned about that), the conversation could shift, and I could just be a regular mom—but one who was *cool* for the mobility equipment I used. I used to glue eyes on my prosthetic foot—so it could "stare back." One kid would sometimes sneakily peel my eyes off as I read (but no big loss, because they come in bulk). One day, I was walking the halls of the school to get to a classroom. I always get stares and turned heads (with whispers of "robot" or "whoa"), and this was a time of day where there were lots of kids in the hall in lines going in different directions. As I was walking by, I could hear one of the friends of my youngest say loudly to his whole class, irritated at their reactions: "That's just Annabelle's mom!" Totally unfazed by my existence and appearance in that hall. I wish we could all be so chill as that kid.

ACKNOWLEDGMENTS

I don't write a lot about my kids or spouse in this book. Some of our stories are theirs, but my kids are not yet old enough to give their fair permission. However, this book would be impossible without their love and support. Thank you, Annabelle, Zephyr, and Randy.

I've been writing this book since 2015—which is to say that I have had the idea for something like this book that has morphed and changed over the course of the last eight years, and also that I have mostly been *not-writing* this short(ish) book, distracted with other writing and teaching and grant applications and two cancer recurrences and many doctors' appointments and kid shenanigans and doing life things. I am grateful for the many friends, family, editors, and colleagues who have cleared my way toward research and writing, and to the team at Norton who has included me in the Norton Shorts series and brought me into the celebration of 100 years of W. W. Norton & Company. It's an incredible honor to be part of this series. Thank you, Alane, Sasha, and Mo.

I owe a big shout-out to my freelance editor/colleague Heath Sledge, who read each of these chapters first, and my writing-expert friend Monique Dufour for giving me confidence and camaraderie in writing. And I need a great deal of peer pressure to write: thank you to my writing group colleagues over a long course here, especially Cora

Olson, Christa Miller, and Ed Gitre, but also many others. Our hijinks keep me going. Thanks to early readers Megan Shew and Rebecca Webb for helpful feedback.

I got a feel for writing for a more general audience for the first time with the history of medicine blog *Nursing Clio*, big thanks to Laura Ansley; and I received excellent advice from David Perry in a writing for the public workshop. The material about tropes in Chapter 3 is rewritten from "How to Get a Story Wrong," published in the new journal *Including Disability* in 2022. Thank you, Stephanie Cork, Ron Padrón, and Sara Olsen, for soliciting this work.

I owe a huge debt to Virginia Tech's Tom Ewing, Karen Roberto, and Yancey Crawford for their grant-writing support. With their encouragement, I applied for a National Science Foundation CAREER grant (#1750260) in 2017, which has supported my research on technology, narrative, and disability—this book is one outcome of that research. I am required now to say: *the opinions represented in this book are not those of the National Science Foundation.* I thank a large group of research assistants and former Tech & Disability students in relation to this grant.

Special thanks to close colleagues and former students through whom I have learned so much, especially Hanna Herdegen, Dylan Wittkower, Robert Rosenberger, Crystal Lee, Ellen Samuels, Emily Ackerman, Joshua Earle, Damien P. Williams, Mahtot Gebresselassie, Manasi Shankar, Alice Fox, Jack Leff, Kristen Koopman, Reem Awni, Martina Svyanek, Liz Spingola, Adrian Ridings, Hannah Jane Upson, Amanda Leckner, and Daniela Pereira. I couldn't exist on campus without the Disability Alliance and Caucus at VT: thanks to Annabelle Fuselier, Lydia Qualls, Phoenix Phillis, Emily Burns, and Elizabeth McLain, and too many others to name!

There are too many people to thank, but I am ever grateful for my fellow board and staff members through my local New River Valley Disability Resource Center and for the vibrant, varied, spread-out community that exists through disability blogs, personal blogs, disability Twitter, critical disability studies Facebook, and, very personally, online cancer and amputee groups. The friends I see mostly online are still very much a part of my life, especially when we're talking at odd hours on prednisone or from hospital rooms, sometimes going through things that are unrelatable to people outside.

NOTES

1   You can buy one here: https://www.teepublic.com/t-shirt/5201256-make-it-accessible?store_id=18557/.
2   I've seen this idea in the writing of Ashe Grey, Jaipreet Virdi, Bill Peace, The Cyborg Jillian Weise, and in conversation with many others.
3   Disability-studies scholar Michelle Nario-Redmond, in her book *Ableism: The Causes and Consequences of Disability Prejudice* (2020), explains the origins of the word "ableism" in the disability rights movement in the USA and Britain "serve as a parallel to sexism and racism." In Nario-Redmond's description there are three components to ableism: affective emotions or attitudes, behavioral actions and practices, and cognitive beliefs and stereotypes.
4   Lewis continues to update their working definition of ableism and provides a longer definition and context here: https://www.talilalewis.com/blog/january-2021-working-definition-of-ableism/.
5   Mallory and I wrote together about our teched-up embodiments with our friend Bethany Stevens, who is not an amputee but also uses different types of mobility equipment: https://catalystjournal.org/index.php/catalyst/article/view/29617.
6   This is true in almost all countries (and in the USA was represented in 2022 in the Public Charge Rule). Especially for people with more documented and more apparent disabilities (where they cannot hide their disability status when applying for citizenship), there are real barriers toward permanent residency and citizenship. This is actually a huge problem for refugee resettlement. Even when there are host nations willing to accept nondisabled refugees, often the people who are not accepted for immigration are disabled. Though the ways in which immigration is hostile or unattainable to disabled people does not often make the news, it did in 2016 when Canada at one point refused the residency of a family where the son had Down syndrome: https://www.cbc.ca/news/canada/toronto/down-syndrome-immigration-1.3492810/.

7   They recently updated this to the less energizing "Claim Your Role," which our local disability group has satirized as "Claim Your Roll" in campaigns for campus accessibility—the addition of curb cuts and more accessible pathways.

8   For more information about Native American conceptions of disability and resources to learn more about these topics, see Marisa Leib-Neri's 2015 overview, "'Everything in Nature Goes in Curves and Circles': Native American Concepts of Disability": https://lewiscar.sites.grinnell.edu/HistoryofMedicine/uncategorized/everything-in-nature-goes-in-curves-and-circles-native-american-concepts-of-disability/.

9   This is not simply historical. More recent accounts and scholarship reflect similar acceptance and social inclusion of disabled community members. For one account of more current interviews with Lakota people, see: Lilah Morton Pengra and Joyzelle Gingway Godfrey, "Different Boundaries, Different Barriers: Disability Studies and Lakota Culture," *Disability Studies Quarterly*, 21, no. 3 (2001): 36–51. Online at: https://dsq-sds.org/article/view/291/331/.

10 Simi Linton, from the documentary about her life: *Invitation to Dance* (2014).

11 For information from #CriptheVote about their use of the "crip" component of the hashtag, see their blog: https://cripthevote.blogspot.com/2018/03/why-we-use-cripthevote.html/.

12 Lizzie Presser gives slightly different and more recent statistics, like: "Despite the great scientific strides in diabetes care, the rate of amputations across the country grew by 50% between 2009 and 2015. Diabetics undergo 130,000 amputations each year, often in low-income and underinsured neighborhoods. Black patients lose limbs at a rate triple that of others." See https://features.propublica.org/diabetes-amputations/black-american-amputation-epidemic/.

13 Data from diagnosis in the wake of Covid-19 suggests that Covid-19 will lead to diabetes in more people, and especially in those who have been infected multiple times. See https://wexnermedical.osu.edu/blog/why-are-people-developing-diabetes-after-having-covid19 and https://www.nature.com/articles/d41586-022-00912-y

14 I wrote a little about this kind of interaction in my 2019 article in the "Nursing Clio" blog: https://nursingclio.org/2019/04/23/stop-depicting-technology-as-redeeming-disabled-people/.

15 Vivian Sobchack has a great article on "Choreography for One, Two, and Three Legs," about some of the possibility and experience of being an amputee dancing in different configurations—on one leg, with a prosthesis, and with crutches. I love how we create.

16 For you philosophy nerds out there, I am definitely playing on William James's ideas about radical empiricism and truth. I read through the work of Davis Baird on thing knowledge; my prior work was about technological knowledge, and there's so much to say about prosthetic devices in this vein! My smart editor friend Heath Sledge put it better in reference to James: "For William James, it's all about process, and what we at the moment take to be the truest thing we have is constantly changing and adjusting to facts. It sounds a little like the process of having a prosthetic—you are always trying to approach that 'it's exactly like my flesh leg was' condition, but you never completely get there, or not for long."

17 Jordan and Jen Lee Reeves coauthored a memoir called *Born Just Right* (New York: Jeter 2019) and have transitioned the old blog into a nonprofit devoted to creative disability-forward design for kids: https://www.bornjustright.org/

18 This is an astute observation made by Mallory Kay Nelson and expanded upon in Mallory Kay Nelson, Ashley Shew, and Bethany Stevens,"Transmobility," *Catalyst* 5, no. 1 (2019), https://doi.org/10.28968/cftt.v5i1.29617.

19 My emphasis added; see Walker's website, Neuroqueer: The Writings of Dr. Nick Walker: https://neuroqueer.com/neurodiversity-terms-and-definitions/.

20 Testimonies about student difficulties with proctor software can be found on Futurism.com: https://futurism.com/college-students-exam-software -cheating-eye-tracking-covid.

21 Not only are autistic people more likely than others to be nonbinary or transgender, but also gender-inclusive accessible restrooms are *great* for people who need attendant care to use the restroom and may have a person of another gender assisting them.

22 There's actually a lot of theorizing about different models of disability. The International Association of Accessibility Professionals has seven different models that are part of its certification test (and I can think of at least three additional models from disability studies that this material doesn't touch). It's not that any one model is right or wrong, but they offer different ways to think about who counts as disabled, where we can make changes or should look to find and address problems, who counts as an authority on disability, and so on.

23 Even the Kennedy family institutionalized and lobotomized Rosemary, sister of John F. Kennedy.

24 I got to see part of this film in progress during early 2020 at University of Virginia, where Snyder and Mitchell answered questions and gave context afterward.

25 These stats are pulled from an ASAN flyer about Autism Speaks that references tax forms for 2018.

26 *The ABCs of Autism Acceptance* (Autonomous Press, 2017) and *No You Don't* (Unstrange, 2013).

27 The terms "autistic meltdown" and "autistic burnout" would also be things autistic adults would add to this list. We might also talk about the higher incidence of eating disorders and alcoholism among autistic people too here, which can be strategies to cope with demands and anxieties produced in coercive contexts.

28 Autistic rhetorician Remi Yergeau has a wonderful article doing autoethnography on how people talk about autistic people in Theory of Mind research. In this piece, they show how "theories about ToM [Theory of Mind] deny autistic people agency by calling into question their very humanity and, in doing so, wreak violence on autistic bodies." For more information about the problems with and limits of theory of mind (ToM) as it is currently constructed in psychology and cognitive studies, see Remi Yergau, "Clinically Significant Disturbance: On Theorists Who Theorize Theory of Mind," *Disabled Studies Quarterly* 33, no. 4 (2013), https://dsq-sds.org/article/view/3876/3405

29 Pavlov was not nice to his dogs, if you check out what he was doing in this and in his later experiments, and he didn't use bells—more likely a buzzer. But the mythology is a bell, and drooling, and not buzzers and invasive cheek surgeries to measure amounts of drool.

30 A list of types of behavioral therapy can be found here: https://www.healthline .com/health/behavioral-therapy#types/.

31 The Living Archive & Repository on the Judge Rotenberg Center's Abuses can be found on Lydia X. Z. Brown's website: https://autistichoya.net/ judge-rotenberg-center/.

32 Thanks for your work, heynowhannah: https://www.tiktok.com/@ heynowhannah

33 November 2020's Choices and Challenges—Technology & Disability: Counternarratives, https://candc.sts.vt.edu.

34 The words "We Move Together" occur to me a lot. And I love the children's book about Disability Justice called *We Move Together* (AK Press, 2021), authored by Kelly Fritsch, Anne McGuire, and Eduardo Trejos.

35 "Dog" post on Reddit's r/ADHDmemes: https://www.reddit.com/r/ ADHDmemes/comments/nmo16u/dog/.

36 See "Land Grab Universities," https://www.landgrabu.org/, which offers an investigative report from *High Country News* by Robert Lee, Tristan Ahtone, Margaret Pearce, Kalen Goodluck, Geoff McGhee, Cody Leff, Katherine Lanpher, and Taryn Salinas that tracks "the ties between indigenous dispossession and the funding of land grant universities."

37 *ProPublica* has done a feature on the plant: https://features.propublica.org/
military-pollution/military-pollution-open-burns-radford-virginia/.

38 Imani Barbarin, Alice Wong, and other social media–wielding disability leaders have been very clear in this messaging since very early in March and April
2020. Covid has taken an incredibly unequal toll on the already-existing disability community, as disabled people were more likely to die from Covid,
or pick up additional disabilities, or live in intense isolation; many of us also
lost access to the routine medical care we are more likely to need as hospitals
filled and in-person care was limited. The National Council on Disabilities
(NCD) offered a 260-page report in 2021 that detailed some of these issues
that occurred in the first year of the pandemic alone: https://ncd.gov/sites/
default/files/NCD_COVID-19_Progress_Report_508.pdf

39 The Hastings Center is an independent bioethics research institute, and
just this year they have had an ongoing series on flourishing, with conversations among invited disabled disability studies scholars. https://
www.thehastingscenter.org/who-we-are/our-research/current-projects/
the-art-of-flourishing-conversations-on-disability-and-technology/.

40 I also thought way too much about the Bene Gesserit in Frank Herbert's
Dune. See my *Medium* article "Water of Life: Chemo and Everything After":
https://medium.com/@ashleyshew/water-of-life-aef9569c62e5/.

41 The schedule was one week staying at the local hospital being infused 24/7,
two weeks at home and feeling sick from that treatment, two weeks in a row
at the far-away hospital for however long it took me to clear enough of the
chemo to their satisfaction (so, usually a weekend home with Monday check-
ins and five days there, but always hopeful to clear more quickly, and often
disappointing myself and sometimes taking longer). We repeated this schedule three times, then I got my big leg surgery (and a bonus surgery after), and
then I had five more rounds of this five-week rotation. We had a whole system
for coming into the house after a hospital stay: spraying all the hospital things
down and sticking everything in the washer, and then usually putting me in
bed, so exhausted from all the night checks and 5 a.m. blood draws and all the
little things that keep a person from actually resting in a hospital, in addition
to being beat up from chemo.

42 I think almost every day about the other young adults I talked to, and became
friends with, though they were far away, who were going through much of the
same. I'm so glad Irene, Florencia, and Autumn are still here, and I wish Oli-
ver, Brittany, and Heather were.

43 The same Emily Satterwhite who locked herself to construction equipment

organized a meal train so that people could deliver meals to our family on a schedule without a lot of extra communication. Both my parents and my spouse's spent stints with the kids during my first of three hospitalizations. Then, my youngest sister took a whole semester off of college to keep things calm and steady for my kids during my amputation and the months of chemo that followed. My spouse slept next to me in many weird foldout-chair/bed situations at hospitals; the hospital where I spent two of my three stays was three hours away, so it wasn't easy for him to do a lot of running back and forth. My department chair (in business-speak, my direct supervisor) told me not to worry and filled out whatever paperwork it took to "stop the clock" on my tenure timing. My colleague Jim took over teaching my class—ironically, a class I had developed on the topic of cyborgs, which I was now becoming.

44 Blogs, videos of recent performances, creative workshops, curricula, and more are available on the Sins Invalid website: https://www.sinsinvalid.org/.

45 Audio and the full version of Marlena Chertock's "On that one-way trip to Mars" are available on the Deaf Poets Society website: https://www.deafpoetssociety.com/marlena-chertock-issue-4/.

46 A Decolonizing Mars participant list is here: https://decolonizingmars.com/attendees/.

47 You must read Chanda Prescod-Weinstein's *The Disordered Cosmos: A Journey into Dark Matter, Spacetime, and Dreams Deferred* (Bold Type Books, 2021).

48 Sammus the Rapper also performed in the event that hosted the Autism Tech Panel from Chapter 6: Choices and Challenges—Technology & Disability: Conternarratives (November 2020), https://candc.sts.vt.edu. She was part of a session called the Cyborg Promenade with poet Travis Lau, and it was the best ever.

49 People get ostomies for a variety of reasons—trauma, intestinal cancers, and diseases like Crohn's where sections of intestines are damaged—and they aren't all that uncommon, even though people don't talk about them very much.

50 And it flows both ways: some disabled people are better for space, and space is better for some disabled people. Friends with some types of chronic pain and people with osteogenesis imperfecta (like Marlena Chertock) talk about what a relief on their body it would be to float and do things without the weight of Earth's gravity.

51 *The Book with No Pictures* (Rocky Pond Books, 2014) by B. J. Novak remains an absolute jam, and we'd have to save it for last because the kids would get so pumped up to read it along with me.

INDEX

Norton Shorts

BRILLIANCE WITH BREVITY

W. W. Norton & Company has been independent since 1923, when William Warder Norton and Mary (Polly) D. Herter Norton first published lectures delivered at the People's Institute, the adult education division of New York City's Cooper Union. In the 1950s, Polly Norton transferred control of the company to its employees.

One hundred years after its founding, W. W. Norton & Company inaugurates a new century of visionary independent publishing with Norton Shorts. Written by leading-edge scholars, these eye-opening books deliver bold thinking and fresh perspectives in under two hundred pages.

Available Fall 2023

Wild Girls: How the Outdoors Shaped the Women who Challenged a Nation by Tiya Miles

Against Technoableism: Rethinking Who Needs Improvement by Ashley Shew

Forthcoming

Mehrsa Baradaran on the racial wealth gap

Ruha Benjamin on imagination

Rina Bliss on the "reality" of race

Merlin Chowkwanyum on the social determinants of health

Daniel Aldana Cohen on eco-apartheid

Brooke Harrington on offshore finance

Justene Hill Edwards on the history of inequality in America

Destin Jenkins on a short history of debt

Quill Kukla on a new vision of consent

Barry Lam on discretion

Matthew Lockwood on a new history of exploration

Natalia Molina on the myth of assimilation

Rhacel Salazar Parreñas on human trafficking

Tony Perry on water in African American culture and history

Jeff Sebo on the moral circle

Tracy K. Smith on poetry in an age of technology

Dennis Yi Tenen on literary theory for robots

Onaje X. O. Woodbine on transcendence in sports